1 MONTH OF
FREE
READING

at
www.ForgottenBooks.com

By purchasing this book you are eligible for one month membership to ForgottenBooks.com, giving you unlimited access to our entire collection of over 1,000,000 titles via our web site and mobile apps.

To claim your free month visit:
www.forgottenbooks.com/free1118603

ISBN 978-0-331-40064-9
PIBN 11118603

This book is a reproduction of an important historical work. Forgotten Books uses state-of-the-art technology to digitally reconstruct the work, preserving the original format whilst repairing imperfections present in the aged copy. In rare cases, an imperfection in the original, such as a blemish or missing page, may be replicated in our edition. We do, however, repair the vast majority of imperfections successfully; any imperfections that remain are intentionally left to preserve the state of such historical works.

Historic, archived document

Do not assume content reflects current
scientific knowledge, policies, or practices.

Breakage Points for Four Corn Translocation Series and Other Corn Chromosome Aberrations

Maintained at the California Institute of Technology

January 1961

Agricultural Research Service
U. S. DEPARTMENT OF AGRICULTURE
in cooperation with
California Institute of Technology

CONTENTS

Prepared by

Crops Research Division
Agricultural Research Service
UNITED STATES DEPARTMENT OF AGRICULTURE

in cooperation with

California Institute of Technology

BREAKAGE POINTS FOR FOUR CORN TRANSLOCATION SERIES AND OTHER CORN CHROMOSOME ABERRATIONS MAINTAINED AT THE CALIFORNIA INSTITUTE OF TECHNOLOGY[1]

Albert E. Longley
Formerly Geneticist, Crops Research Division, Agricultural Research Division

INTRODUCTION

A cooperative project was set up by the Agricultural Research Service with the California Institute of Technology shortly before the Bikini bomb was exploded in 1946 ("Operation Crossroads"). The cooperative project has isolated and maintained a large number of translocations and other chromosomal rearrangements from corn material treated with ionizing radiations.

The cooperative arrangement is now being terminated after 13 years of intensive study. Dr. E. G. Anderson of the Institute is preparing the translocation and inversion stocks for a seed pool that can be drawn from by corn geneticists and cytogeneticists as needs for these radiation-induced markers materialize.

Each translocation and inversion stock is characterized by certain cytological features. These features pertain to the chromosome or chromosomes involved and the position of the breakages induced by the various ionizing radiations. An accurate placement of breakage points is the aim in cytological studies involving translocations and inversions.

The data presented have resulted from studies that were repeated whenever successive generations of a translocation or inversion became available. Breakage point positions established by the use of only a single plant are not well fixed; therefore a seed stock carrying a certain arrangement of chromosome break positions may be missing from the seed pool being organized.

Originally this project concentrated on detecting the presence, determining the character, and comparing the prevalence of the chromosomal rearrangements produced by ionizing radiations from different sources. As time passed the emphasis shifted to the isolation and maintenance of translocations and inversions, irrespective of source of irradiation. It is planned that these translocations and inversions will constitute a nucleus to which other radiation-produced markers may be added.

Essential data are herein reported to make this initial collection useful to others. The report supplements data of an earlier publication of the Crops Research Division[2] by the addition of two translocation series and the inclusion of a list of the numerous inversions detected and isolated. All translocation data are arranged in order in a summary table according to the chromosomes involved and the breakage point position along each chromosome arm.

[1] Cooperative investigation between the Crops Research Division, Agricultural Research Service, U.S.D.A., and the California Institute of Technology, Pasadena, Calif.
[2] Longley, A. E. Breakage points for two corn translocation series maintained at the California Institute of Technology. U. S. Dept. Agr., Agr. Res. Serv. ARS 34-4, 31 pp, 1958.

MATERIALS AND METHODS

It was planned to subdivide the data not ready for publication in 1958 into three addi_tional series. This plan has been modified, and the first series of this report (series 3), involving irradiation of corn seed, combines the data from several lesser series. The second series (series 4) tabulates the data for altered chromosomes induced by treating corn pollen with 1,000 r. X-ray exposures.

The following tabulation will serve as a key to tie symbol numbers to a definite ionizing radiation treatment and group.

Seed Treated

Miscellaneous X-ray treat- ments (early) --------	All symbol numbers with letters. (Lower-case letters indicate permanent symbols)
X-ray 5,000 r. ----------	5068 to 5098; 5156 to 5179
X-ray 10,000 r. ---------	5188; 6067 to 6061; 6178 to 6200; 6673 to 6697; 7036 to 7043.
X-ray 15,000 r. ---------	4394 to 4398; 4456 to 4484; 4573 to 4597; 4662 to 4698; 4773 to 4791; 4861 to 4898; 4963 to 4997; 5253 to 5300; 5355 to 5386; 5752 to 5800; 5863 to 5896; 5951 to 5982; 6261 to 6293; 6353 to 6373; 6462 to 6498; 6557 to 6600; 6653 to 6671.
X-ray 20,000 r. ---------	5391 to 5393; 5453 to 5499; 5557 to 5597; 5651 to 5693; 7067 to 7088.
X-ray 25,000 r. ---------	7096 to 7162.
Bikini Lot 1 -----------	5013 to 5045; 5131 to 5144.
Bikini Lot 2 -----------	6018 to 6027; 6128 to 6131.
Bikini Lot 3 -----------	4308 to 4384; 4401 to 4447; 4505 to 4545; 4613 to 4643; 4711 to 4759; 4801 to 4884; 4932 to 4937; 5201 to 5250; 5304 to 5343; 5407 to 5441; 5512 to 5539; 5602 to 5648; 5704 to 5724; 5803 to 5831; 5906 to 5949; 6222 to 6225; 6323 to 6349; 6401 to 6439; 6504 to 6534; 6612 to 6623; 6722 to 6943.
Bikini Lot 6 -----------	6103; 6979 to 6981; 7006.
Bikini Lot 7 -----------	6994.
Eniwetok E-2 -----------	8001 to 8010; 8041 to 8129; 8163 to 8186; 8249 to 8302; 8321 to 8397.
Eniwetok E-3 -----------	8023 to 8027; 8143 to 8145.
Oak Ridge 15,000 r. gamma -	8032 and 8219.
" " 20,000 r. " -	8405 to 8428; 8483 to 8491.
" " 40,000 r. " -	8439 to 8465.
Miscellaneous radiation treatments (late) ------	4301-39 to 4307-12; 7205 to 7535; 8513 to 9035.

Pollen Treated

X-ray 1,000 r. ----------	001-3 to 073-8

PRESENTATION

The method of calculating breakage point positions was described previously (1958). Some positions where both breaks in a translocation are near the centromeres (ctr.) present a special problem. It is difficult to place these breakage points definitely as to the arms involved. Nonhomologous pairing in these types of configuration frequently places the assumed breakage points in the wrong arms in both chromosomes, since the opening used to detect the position of the breakage points may appear on either side of the centromere.

Nonhomologous pairing in heterozygous translocation configurations of the type described in the previous paragraph can be traced when chromosome 6 with its break in the organizer is involved with a second chromosome having its break the same distance from the centromeres as the break in 6. The heterochromatic area adjacent to the organizer serves to mark the position taken by the major portion of the short arm of chromosome 6.

Figure 1 illustrates homologous and nonhomologous pairing in a 6-9 (F4778) translocation, and such pairing in a similar diagram of a 7-8 translocation without an identifying marker. The microscope will not detect the presence of nonhomologous pairing in the latter type of translocation.

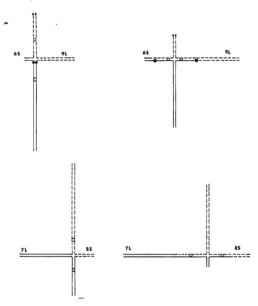

Figure 1.--Homologous and nonhomologous pairing in two translocations. A, Homologous pairing in a 6-9 translocation; B, nonhomologous pairing in this 6-9 translocation. C, Homologous pairing in a 7-8 translocation; D, nonhomologous pairing in this 7-8 translocation.

Translocations

Tables 1 to 3 present new data on breakage point positions from corn material. These data when combined with tables 1 and 2 of the 1958 publication list all translocations. They are arranged according to their symbols and according to the position of their breaks.

A few of the earlier listed translocations were omitted and a few were restudied and corrected. These are given in table 1 of this report. Data for a group of corn translocations originating from several seed sources and different seed treatments make up series 3 (table 2). The data for a group of translocations originating from corn pollen exposed to 1,000 r. X-ray (series 4) are given in table 3.

The cooperative project has produced an appreciable number of corn translocations (1,003). From them one is able to draw a translocation combining any two chromosomes and with breakage points close to almost any desired region. It is from table 4--which presents summary data for all the translocations as a group according to the position of their breaks--that translocations involving specific chromosome combinations with breakage points at specific positions may be selected most readily.

Translocation data, to be used for comparative study, should be taken from a random sample. This randomness should be characteristic of the whole population and of the subdivisions of the population. The translocations of this report, with a few possible exceptions, are thought to have come from a random population.

One of the factors possibly contributing to a nonrandom population of translocations is the failure to detect translocations originating from chromosome breaks near the distal end of chromosome arms. Such translocations, when in the heterozygous state, produce little or no visible pollen sterility. Consequently translocations of this type are lost from a population, because detection and maintenance of a translocation depend upon the presence of visible pollen sterility. However, as this loss of translocations is considered to be nonselective for any particular chromosome, the translocations presented in table 4 may be used comparatively in tables.

A second factor that might contribute toward the production of a nonrandom population of translocations is preferential selection. If, for example, the detection and preservation of stocks having blue fluorescence is given preferential treatment, the 9L translocations involving breaks well out on the long arm of 9 will be increased, because these translocations, when heterozygous, are often characterized by having blue fluorescent anthers. The data of table 4 do not suggest any appreciable increase of translocations with breaks in the distal part of 9L.

All translocations that in the heterozygous condition were readily detected by pollen sterility were retained, and no serious bias toward translocations linked to a particular gene or group of genes existed during these studies. It is assumed, therefore, that the data of table 4 and subsections within table 4 are not far from representing random populations.

Two measures were selected for use in comparing the data from different radiation treatments, from different chromosomes, and from different parts of the various chromosomes and chromosome arms. One of these measures gives the mean position of breakage points in chromosome arms. The other measure gives the density of breakage points. This density is expressed as the ratio of observed number of breaks divided by expected number of breaks. When the total number of breaks is known for the total chromosome length[3] under consideration, the expected number of breaks for any fraction of this chromosome length can be calculated and compared with the observed. For example, the total number of breaks given in table 4 is 2,006; the total chromosome length is 552.6μ. Therefore, the average (expected) density of breaks per micron is 3.62. From this figure the expected number of breaks in the 199.8μ in the 1μ to 10μ section adjacent to the centromeres and in the 8μ in the distal 40μ to 50μ section are calculated. The first line of table 5 evidences that the observed numbers of breaks in these two sections are respectively 1.43 and 0.4 times as great as the calculated break densities for the two sections.

Tables 5 and 6 have been prepared to show breakage point positions and densities for the total translocation population (table 5) and three subpopulations (table 6).

Mean breakage point position seems to depart significantly from 0.5, the midpoint for chromosome arms. This shift in the mean position toward the centromere must be due in part to failures in detection and in part to losses in maintaining those translocations involving breakage points at distal positions of the chromosome arms.

---

[3] For a discussion of chromosome and chromosome arm lengths, refer to A. E. Longley, Knob positions in corn chromosomes. Jour. Agr. Res. 59: 475-490. 1939.

For most areas adjacent to the centromeres, the density of breaks that entered into production of translocation is much above that expected. This observation has led to the conclusion that the more densely staining areas of a chromosome thread are associated with a high concentration of breaks. Thus the deeply staining areas adjacent to the centromeres help to shift the mean position of breaks proximally.

Staining intensity of the chromosome thread and not the proximity of the centromere appears to be the factor associated with increases in density of breaks. An increase in the density of breaks in a nonproximal section is apparent in the study of the data for 7L in table 5. This knob-bearing section has a high density while the section just proximal to it has a low density of breaks.

It seems clear from the data of table 5 that not only the position of breaks but also the density of breaks is influenced by the morphological features of chromosomes, chromosome arms, and sections of arms.

The data of table 6 were selected with the idea that the three different radiation treatments might have acted differently in producing breaks in the same chromosome arm or in different arms with similar morphological characteristics. The first line of the table suggests that the 1,000 r. X-ray treatment of pollen caused the mean breakage points to depart most and the Bikini treatment of seed caused the mean breakage points to depart least from the midpoint of the chromosome arms.

Pollen treated with 1,000 r. X-ray shows, in general, the least and the seed treated with 15,000 r. X-ray the greatest variation around their mean break position and mean break density.

The three radiation treatments show distinct differences between some chromosome arms in both the position and the density of breaks. The differences are least for arms taking a uniform stain, whether it be heavy or light, throughout most of their length. The differences in position and density increase as the lack of uniformity in staining increases.

One may therefore conclude that breaks occurring in heterochromatic areas of a chromosome are utilized more frequently in translocation formation than breaks in euchromatic areas, and that different radiation treatments affect differently the productions of breaks or the utilization of breaks in translocation formation.

Inversions

The number of inversions (60) found in the plants grown from treated material is considerable. It seems likely that the usual cytological studies do not detect short inversions, which seldom produce a loop configuration at mid-prophase.

The breakage point positions for the inversions detected during this study are given in table 7. Dr. Anderson has not been able to give this group of chromosome alterations the same attention as has been given to the translocation group. He is endeavoring, however, to make seed stocks of many of the inversions of table 7 available to corn cytogeneticists and geneticists.

Deletions

Deletions in corn chromosomes are rarely transmitted through the gametophyte phases. Consequently they were observed more frequently in data obtained in 1946 from progenies grown directly from seeds exposed that season, and later in data from F_1 progenies derived from pollinations made with X-ray exposed pollen.

The seed treated material gave 8 deletions among the 36 chromosome alterations studied. The pollen treatment gave 23 deletions among the 117 plants containing chromosome alterations. One exceptional deletion has been transmitted successfully (#4974). This deletion was of the proximal 75 percent of the short arm of chromosome 10. It included enough of the centromere so that it behaved as a short chromosome; when it was accompanied in a gamete by the complementary part of chromosome 10, the gamete was viable and the two chromosomes derived from chromosome 10 were transmitted to later generations.

SUMMARY AND CONCLUSIONS

Chromosomal rearrangements, such as translocations, inversions, and deletions were induced in corn seed and pollen by several types of ionizing radiation during this study. It is planned that seed stocks of many of the 1,003 translocations and 60 inversions will be available to corn geneticists and cytogeneticists.

Analysis of the 1,003 translocations has produced data that can be summarized as follows:

The chromosome breaks that unite to produce transmissible translocations are not distributed at random among the 10 chromosomes, among the 20 chromosome arms, nor among the 10μ sections of the 20 arms. This nonrandomness is present regardless of the source of ionizing radiation or the material treated.

The highest concentration of breaks is found in the more densely staining areas of the mid-prophase chromosomes. Thus, the concentration of breaks in an arm or a portion of an arm may be a measure of the staining density of the arm or a fraction of the arm.

Arms with a uniform staining density over much of their length, whether heavy or light, are affected the same by different types of ionizing radiation. On the other hand, arms with areas staining nonuniformly light and heavy show, with different doses of radiation, variation in position and density of the break bringing about the translocation.

In the comparison of three radiation treatments (seed exposed to 15,000 r. X-rays, Bikini exposed Seed Lot 3, and pollen exposed to 1,000 r. X-rays), it is determined that the 1,000 r. X-ray treatment shows the greatest, the Bikini Lot 3 the least, tendency toward producing breaks near the centromere in the production of translocations. The 15,000 r. X-ray treatment of seed shows the greatest variability around the mean in the break position and in the concentration of breaks.

In conclusion, it can be stated that a relationship exists between staining ability of the chromosome threads and the production of translocations. Differences in this relationship result in differences both in the position and in the density of the breaks that are associated with translocations produced in treating a material either with different intensities of ionizing radiation of one type or with similar types of radiation at a uniform intensity.

1 chromosome translocations, and corrections to the mean position of
aks, for series 1 and 2 given in tables 1 and 2, respectively, of the
ARS 34-4, 1958.

SERIES 1

Translocation

1-2	1L.21	2L.37	1	3
2-3	2L.79	3S.90	2	5
9-10	9S.21	10L.60	1	2
5-9	5L.42	9L.15	3	3
4-9	4L.41	9L.93	3	9
4-8	4S.40	8L.14	1	4
7-9	7L.60	9ctr.	2	5
1-7	1L.12	7L.24	3	9
1-4	1S.03	4S.04	5	9
3-6	3L.85	6S.80	2	4
5-7	5S.20	7L.14	4	10
1-2	1L.43	2L.50	2	6
1-5	1L.15	5S.19	4	5
4-7	4S.31	7L.66	4	5
1-6	1S.40	6L.60	3	5
7-10	7S.78	10L.88	2	4
6-7	6L.29	7L.45	1	6

>n.

TABLE 2.--The mean position for the breaks in the chromosome translocations from
several radiation treatments--series 3.

Symbol	Translocation	Mean position of breaks		Source of data--	
		Ratio	Ratio	Plants, No.	Drawings, No.
7402------------------	6-7	6L.97	7L.14	1	4
7464------------------	3-10	3S.47	10L.60	1	3
7509------------------	1-8	1L.28	8L.21	8	13
7535------------------	1-9	1S.33	9S.27	3	7
8001------------------	1-9	1S.51	9L.24	4	9
8004------------------	4-8	4S.27	8L.84	2	5
8006------------------	3-7	3L.88	7L.90	6	11
8006------------------	4-5	4S.37	5S.18	2	6
8023------------------	3-8	3L.18	8L.16	2	8
8027------------------	2-4	2L.15	4L.43	2	8
8032------------------	3-9	3S.26	9L.96	3	8
8041------------------	1-5	1L.80	5L.15	7	14
8045------------------	2-7	2S.12	7L.06	1	3
8048------------------	1-3	1L.11	3S.18	2	7
8069------------------	4-5	4S.34	6S.71	3	10
8103------------------	4-7	4S.81	7L.76	3	11
8104------------------	3-5	3L.05	5L.08	1	2
8108------------------	4-5	4S.37	5S.72	1	3
8129------------------	1-9	1S.53	9L.27	3	9
8143------------------	6-7	6L.35	7L.36	4	11
8145------------------	3-6	6L.17	6L.26	2	4
8163------------------	4-8	4S.17	8L.86	1	4
8219------------------	2-10	2L.50	10L.35	3	11
8219------------------	5-6	5L.71	6S.84	6	8
8249------------------	1-4	1L.26	4L.63	2	6
8265------------------	1-9	1S.58	9L.27	2	9
8302------------------	1-9	1S.55	9L.29	5	15
8321------------------	2-5	2L.86	5L.11	4	9
8322------------------	2-7	2L.76	7L.74	1	3
8339------------------	4-6	4L.87	6L.79	7	16
8345------------------	5-10	5L.87	10S.61	5	11
8346------------------	7-8	7S.49	8S.30	1	6
8347------------------	1-5	1S.84	5L.52	3	10
8349------------------	3-10	{ 3S.50	10S.40 }	5	17
＊		3S.26	10L.34		
8350-------------·----	3-8	3L.75	8S.60	4	5
8351------------------	3-5	3L.75	5L.68	4	11
8367------------------	3-8	3S.28	8S.52	3	6
8368------------------	1-4	1S.14	4S.30	1	3
8374------------------	4-7	4L.24	7L.55	1	3
8375------------------	1-10	1L.69	10L.64	3	12
8376------------------	2-8	2L.95	8L.03	4	13
8379------------------	5-6	5L.89	6L.67	3	6
8380------------------	4-6	4S.47	6L.18	5	11
8383------------------	7-9	{ 7S.20	9L.35 }	6	20
		7L.13	9S.31		
8386------------------	5-9	5L.87	9S.13	9	16
8388------------------	1-5	1L.30	5S.25	5	12
8389------------------	1-9	1L.74	9L.13	12	22

TABLE 2.--Continued.

Symbol	Translocation	Mean position of breaks Ratio	Ratio	Plants, No.	Drawing No.
8395------------------	4-5	4L.63	5L.82	4	12
8397------------------	3-4	3S.74	4S.55	1	4
8405------------------	1-3	1L.60	3L.31	5	10
8407------------------	2-4	2ctr.	4ctr.	1	2
8412------------------	3-10	3S.39	10S.36	2	8
8415------------------	1-6	1L.29	6S.82	3	8
8420------------------	5-8	5S.90	8L.33	4	13
8428------------------	2-8	2L.16	8L.10	3	6
8428------------------	4-6	4L.32	6L.28	2	8
8439------------------	6-9	6L.06	9S.73	5	10
8441------------------	2-6	2L.94	6S.79	3	8
8443------------------	3-4	3L.12	4L.13	3	5
8447------------------	3-9	3S.44	9L.14	5	13
8452------------------	1-6	1S.80	6L.52	5	13
8456------------------	4-8	4S.22	8L.75	13	25
8457------------------	5-9	5L.78	9S.83	2	10
8458------------------	2-8	2S.22	8L.32	8	8
8458------------------	5-8	5S.68	8L.32	8	12
8460------------------	1-9	1S.13	9L.24	6	13
8465------------------	2-7	2L.27	7S.56	1	3
8465------------------	3-9	3S.27	9L.41	6	14
8465------------------ 8491------------------ }	1-10	1L.45	10L.76	22	40
8483------------------	2-3	2L.14	3L.12	5	15
8491------------------	5-7	5ctr.	7ctr.	6	8
8513------------------	5-8	5S.34	8L.24	9	18
8525------------------	8-9	8L.06	9S.63	12	27
8528------------------	3-5	3L.06	5L.72	2	7
8536------------------	6-9	6L.18	9S.81	6	8
8541------------------	4-10	4S.45	10ctr.	2	4
8541------------------	3-7	{ 3L.15 3S.35	7S.84 7L.23 }	} 1	4
8553------------------ 8556------------------ }	2-3	2L.26	3L.23	15	22
8558------------------	7-9	7S.22	9L.16	1	4
8562------------------ 8568------------------ 8572------------------ 8574------------------ }	3-9	3L.65	9L.22	9	21
8563------------------	1-4	1L.39	4S.21	4	7
8580------------------ 8583------------------ }	7-8	{ 7S.44 7L.13	8S.41 8L.11 }	} 3	9
8590------------------ 8593------------------ }	5-6	5S.29	6L.25	8	20
8591------------------	5-9	5S.09	9L.25	2	4
8591------------------ 8593------------------ }	4-6	4L.17	6L.24	2	8
8592------------------	1-4	1L.43	4L.69	3	17
8592------------------	6-9	6L.19	9S.41	1	3

TABLE 2.--Continued.

Symbol	Translocation	Mean position of breaks		Source of data--	
		Ratio	Ratio	Plants, No.	Drawings, No.
8598------------------- 8602-------------------	} 1-4	1S.41	4L.81	5	13
8603------------------- 8606------------------- 8607------------------- 8608-------------------	} 4-8	4S.42	8L.35	6	14
8605------------------- 8609-------------------	} 1-6	1S.79	6L.59	5	13
8611-------------------	7-9	7L.23	9L.18	3	8
8622-------------------	4-5	4L.30	5L.52	1	3
8628-------------------	1-2	1ctr.	2L.49	6	11
8628-------------------	2-3	2L.54	3L.44	1	3
8630-------------------	5-7	5L.38	7L.24	2	3
8630-------------------	9-10	9S.28	10L.37	3	10
8634-------------------	3-4	3S.71	4L.75	3	7
8636------------------- 8649-------------------	} 4-9	4L.94	9S.09	5	9
8637-------------------	1-3	1L.37	3S.50	3	4
8640-------------------	1-8	1L.11	8L.16	1	3
8645-------------------	6-10	6L.21	10L.28	3	9
8650------------------- 8658-------------------	} 1-6	1L.79	6L.91	4	9
8651-------------------	6-10	6L.27	10L.28	1	3
8659-------------------	7-9	7S.55	9S.35	3	5
8661-------------------	8-9	8S.66	9L.92	15	29
8662-------------------	2-3	2S.78	3L.83	3	6
8663------------------- 8664-------------------	} 1-4	1S.09	4S.36	6	14
8665-------------------	5-6	5L.58	6L.25	2	5
8666------------------- 8667------------------- 8670-------------------	} 3-8	3S.30	8L.14	3	7
8671-------------------	5-7	5L.96	7L.67	5	6
8672-------------------	3-6	3L.47	6L.87	6	13
8677-------------------	4-8	4S.24	8L.79	2	5
8679-------------------	5-7	5S.09	7S.26	3	7
8682-------------------	2-4	2S.70	4S.25	1	3
8683-------------------	1-8	1L.11	8ctr.	4	9
8696-------------------	5-6	5L.89	6S.80	4	11
8704-------------------	5-9	5L.35	9L.85	12	22
8746-------------------	5-8	5S.84	8L.25	2	4
8764-------------------	4-6	4L.32	6L.90	1	4
8768-------------------	6-9	6L.89	9S.61	2	5
8770-------------------	1-10	1L.09	10L.38	2	4
8782-------------------	1-5	1ctr.	5ctr.	3	6
8786-------------------	2-6	2S.90	6S.77	1	3
8796-------------------	5-8	5L.76	8L.11	2	7
8806-------------------	5-8	5L.72	8S.59	2	4
8818-------------------	5-6	5S.91	6L.93	1	3
8824-------------------	1-9	1S.06	9L.83	1	5

nued.

Translocation

--------	5-9	5S.33	9S.36	1	3
--------	2-3	2S.08	3L.24	5	10
--------	2-10	2S.10	10L.76	3	8
--------	2-4	2S.52	4L.27	1	4
-------- }	1-9	1L.33	9L.23	4	10
--------	4-5	4ctr.	5ctr.	3	7
--------	5-9	5L.37	9L.17	1	3
--------	6-10	6L.51	10L.83	3	7
--------	6-9	6L.27	9L.59	2	4
--------	1-9	1S.21	9L.20	2	8
--------	1-8	1S.53	8L.44	1	3
--------	4-6	4L.70	6L.18	3	5
--------	5-9	5L.43	9L.80	8	20
--------	8-9	8L.13	9L.77	2	8
--------	4-5	4L.17	5S.32	1	2
--------	5-8	5L.17	8L.83	1	2
--------	3-6	3S.23	6L.14	6	11
--------	3-4	3S.75	4L.75	2	5
--------	1-5	1S.56	5S.29	2	5
--------	4-8	4S.58	8L.76	1	5
--------	1-3	1S.49	3L.06	1	3
--------	5-8	5L.16	8L.08	1	4
-------- }	2-6	2L.57	6L.50	4	8
--------	8-10	8L.13	10S.50	1	4
--------	1-9	1L.26	9L.83	6	12
-------- }	4-10	4S.57	10L.89	2	6
--------	1-5	1L.19	5L.76	1	3
--------	1-9	1S.51	9L.24	1	3
--------	4-7	4S.32	7L.64	6	14
--------	2-8	2L.84	8L.68	4	7
--------	2-7	2L.22	7S.47	5	15
--------	1-7	1L.12	7L.25	7	17

Table 3.--The mean position for the breaks in the chromosome translocations from 1,000 r X-ray pollen treatment--series 4.

Symbol	Translocation	Mean position of breaks		Source of data--	
		Ratio	Ratio	Plants, No.	Drawings, No.
001-3-------------------	1-10	1L.86	10L.48	4	8
001-5-------------------	8-10	8L.30	10S.57	3	8
001-13------------------	1-8	1S.39	8L.67	3	7
001-15------------------	2-6	2S.72	6S.87	2	4
001-15------------------	3-7	3S.38	7L.30	9	14
002-12------------------	4-5	4S.29	5S.36	2	8
002-16------------------	2-5	2L.23	5L.35	2	9
002-17------------------	5-8	5L.11	8L.28	1	2
002-19------------------	1-4	1S.88	4L.42	1	3
003-2-------------------	3-6	3L.35	6L.15	1	2
003-5-------------------	2-8	2L.19	8S.72	1	3
003-16------------------	4-6	4L.50	6S.90	6	11
004-3-------------------	7-8	{ 7S.74 / 7L.11	8L.19 / 8L.37 }	2	5
004-7-------------------	3-7	3S.38	7L.26	8	19
004-7-------------------	4-9	4L.28	9L.26	8	15
004-11------------------	1-2	1S.13	2S.36	3	6
004-12------------------	4-8	4L.10	8L.10	1	3
004-13------------------	2-4	2L.14	4S.51	3	7
004-14------------------	1-5'	1S.75	5L.33	1	2
004-17------------------	5-6	5L.60	6L.24	4	10
005-7-------------------	1-8	1L.22	8L.78	6	12
005-14------------------	2-3	2L.12	3S.29	1	3
006-7-------------------	4-5	4L.43	5S.25	3	7
006-10------------------	2-8	{ 2S.31 / 2L.10	8S.64 / 8S.13 }	4	6
006-11------------------	5-10	5L.49	10L.52	1	3
006-17------------------	3-4	3L.10	4S.45	1	3
007-1-------------------	1-4	1L.04	4L.06	1	4
007-8-------------------	3-8	3L.14	8L.11	1	2
007-17------------------	5-8	5L.32	8S.47	3	9
007-18------------------	4-5	4L.22	5L.39	1	3
007-19------------------	1-10	1S.08	10L.08	6	12
008-3-------------------	7-9	7L.80	9L.85	1	3
008-5-------------------	2-6	2S.40	6L.26	1	6
008-16------------------	4-7	4L.86	7L.17	11	16
008-16------------------	7-9	7L.51	9L.62	11	16
008-17------------------	1-8	1S.16	8L.20	3	6
008-18------------------	5-9	5L.29	9L.26	16	28
009-19------------------	2-5	2S.29	5L.11	1	2
010-4-------------------	2-4	2L.13	4S.31	11	22
010-10------------------	2-3	2S.17	3L.13	1	3
010-12------------------	1-7	1S.35	7L.57	2	3
010-17------------------	2-5	2S.19	5L.21	1	4
011-6-------------------	2-3	2L.77	3L.48	1	4
011-7-------------------	2-4	2L.53	4L.76	3	8
011-9-------------------	2-10	2L.39	10L.76	1	5
011-11------------------	6-7	6L.29	7L.29	9	15
011-16------------------	4-6	4S.31	6L.33	4	9
011-20------------------	2-8	2S.58	8L.28	2	4

12

TABLE 3.--Continued.

Symbol	Translocation	Mean position of breaks		Source of data--	
		Ratio	Ratio	Plants, No.	Drawings. No.
012-10-----------------	1-7	1L.91	7L.90	1	3
012-10-----------------	3-8	3L.39	8L.23	1	4
012-16-----------------	3-4	3S.27	4S.30	4	11
013-3------------------	5-7	5S.36	7S.35	3	6
013-8------------------	6-7	6L.31	7L.72	11	17
013-9------------------	1-3	1ctr.	3ctr.	3	9
013-9------------------	5-9	5L.51	9L.82	2	5
013-11-----------------	5-8	5S.59	8S.63	2	7
013-17-----------------	2-8	2S.89	8L.61	2	9
014-5------------------	5-8	5L.19	8L.18	3	9
014-11-----------------	2-4	2L.81	4S.16	9	14
014-11-----------------	2-6	2L.81	6L.20	9	14
014-12-----------------	2-3	2S.43	3L.51	4	9
014-17-----------------	7-8	7L.18	8S.30	2	4
015-3------------------	2-5	2L.16	5S.69	10	15
015-9------------------	1-10	1L.67	10S.46	1	4
015-10-----------------	5-9	5L.50	9L.20	2	7
015-12-----------------	7-10	7ctr.	10ctr.	6	18
015-15-----------------	1-4	1L.49	4S.40	1	3
015-17-----------------	3-8	3S.12	8S.13	1	4
015-18-----------------	1-2	1L.26	2L.16	1	4
015-18-----------------	3-5	3L.18	5S.21	1	3
016-6------------------	4-8	4L.42	8S.45	1	4
016-15-----------------	7-8	{ 7S.63 / 7L.04	8S.52 / 8L.08 }	7	11
016-17-----------------	3-6	3S.48	6L.30	3	9
016-19-----------------	2-4	2L.81	4S.79	1	3
017-3------------------	1-2	1L.67	2S.58	4	11
017-8------------------	5-8	5S.09	8S.25	1	1
017-10-----------------	1-7	1L.82	7L.12	1	3
017-14-----------------	6-8	6S.92	8L.83	1	3
017-14-----------------	6-9	6S.80	9L.50	4	11
017-18-----------------	2-4	2L.39	4L.19	4	7
018-3------------------	2-4	2S.38	4L.47	5	15
018-4------------------	4-5	4L.61	5L.67	7	13
018-5------------------	1-5	1S.53	5L.52	3	9
018-12-----------------	6-9	6L.21	9L.25	5	22
018-12-----------------	8-9	8L.52	9S.49	10	10
018-18-----------------	1-2	1L.16	2L.44	2	6
019-1------------------	2-5	2S.67	5S.51	3	9
019-3------------------	7-10	7L.17	10L.47	5	8
020-5------------------	3-9	3ctr.	9ctr.	16	21
020-7------------------	5-9	5ctr.	9ctr.	8	20
020-19-----------------	1-8	1L.11	8L.38	4	10
021-1------------------	7-8	7L.72	8L.49	5	14
021-3------------------	4-5	4L.62	5S.71	5	8
021-5------------------	4-10	4L.34	10L.33	3	8
021-7------------------	1-3	1S.72	3L.41	5	6
021-7------------------	1-7	1S.22	7S.56	5	6
022-4------------------	2-7	2S.30	7L.24	2	6
022-11-----------------	1-5	1L.27	5L.56	2	5

TABLE 3.--Continued.

Symbol	Translocation	Mean position of breaks		Source of data	
		Ratio	Ratio	Plants, No.	Drawings, No.
022-11-----------------	5-9	5S.30	9L.27	5	9
022-15-----------------	7-10	7ctr.	10ctr.	4	9
022-20-----------------	5-10	5L.65	10S.62	3	8
023-2------------------	2-3	{ 2L.09 2S.14	3S.13 3L.12 }	4	8
023-5------------------	2-3	2S.80	3L.70	2	5
023-13-----------------	5-7	5L.12	7L.15	4	11
023-15-----------------	2-5	2L.28	5S.30	3	6
023-15-----------------	8-10	8S.42	10S.31	2	5
024-1------------------	6-8	6L.42	8L.74	4	9
024-5------------------	1-5	1S.09	5L.89	4	5
024-7------------------	1-9	1S.71	9L.13	4	11
024-11-----------------	3-8	3S.65	8L.49	2	7
024-14-----------------	1-3	1L.27	3S.49	3	5
024-16-----------------	4-10	4L.75	10L.18	3	6
025-4------------------	2-5	2S.26	5S.78	3	5
025-12-----------------	4-6	4S.44	6L.34	7	12
026-2------------------	1-8	1L.49	8L.80	6	9
027-4------------------	2-6	2S.34	6L.11	4	10
027-6------------------	6-7	6L.66	7L.97	4	9
027-9------------------	7-9	7L.61	9S.18	3	6
027-10-----------------	4-5	4L.61	5L.79	6	13
027-17-----------------	4-7	4L.17	7L.31	4	12
028-13-----------------	1-6	1S.56	6L.54	3	7
028-17-----------------	1-2	{ 1L.05 1S.04	2L.08 2S.08 }	6	12
029-3------------------	3-7	3L.11	7L.13	3	6
030-1------------------	5-8	5L.48	8L.78	3	8
030-2------------------	3-9	3S.39	9L.30	6	13
030-8------------------	3-6	3S.27	6S.81	5	10
031-7------------------	2-8	2L.30	8S.44	2	8
031-18-----------------	5-10	5S.58	10S.55	3	8
032-3------------------	3-6	3S.41	6S.78	4	8
032-8------------------	5-9	5L.19	9L.70	2	6
032-9------------------	2-5	2L.40	5S.31	3	8
032-13-----------------	7-9	7L.82	9L.88	8	16
033-4------------------	2-3	2L.27	3L.23	1	4
034-11-----------------	3-9	3L.46	9S.36	10	12
034-11-----------------	8-9	8L.14	9L.18	10	12
034-17-----------------	7-8	7L.05	8S.59	1	4
034-19-----------------	8-10	8L.24	10L.28	3	9
035-2------------------	2-10	2L.85	10L.19	4	7
035-3------------------	6-7	6S.80	7L.20	3	11
035-10-----------------	1-9	1L.89	9S.67	1	4
036-4------------------	1-8	1L.18	8L.59	4	16
036-7------------------	1-2	1S.37	2L.33	3	9
036-15-----------------	3-10	3L.48	10L.64	6	17
036-16-----------------	4-8	4S.66	8L.69	3	7
037-5------------------	2-8	2L.95	8L.54	3	7
037-9------------------	3-4	3L.10	4L.14	4	6
038-8------------------	7-8	7L.52	8L.46	2	7

14

Translocation

4-6	4L.78	6L.29	3	6
2-7	2L.75	7S.68	2	5
3-5	3L.13	5L.14	3	7
1-4	1L.14	4S.26	3	9
5-6	5S.48	6S.82	3	8
1-5	1S.17	5L.61	1	2
9-10	9L.69	10L.92	7	17
9-10	9L.70	10L.90	7	19
1-2	1L.27	2S.57	4	8
6-9	6L.36	9L.36	3	10
8-9	8L.17	9S.34	4	9
2-10	2S.89	10L.40	5	14
3-8	3L.02	8S.40	2	8
1-5	1S.10	5L.63	2	4
4-5	4L.38	5L.33	4	5
5-9	5L.33	9L.55	4	7
6-10	6L.48	10L.51.	4	8
1-5	1L.05	5S.83	4	10
3-10	3L.77	10L.72	5	13
5-8	5L.08	8L.13	3	10
1-2	1L.16	2S.30	2	7
2-8	2L.83	8L.74	6	15
2-8	2L.62	8L.48	4	8
2-4	2ctr.	4ctr.	3	9
4-7	4L.27	7L.25	3	6
2-4	2L.75	4L.66	3	7
5-8	5L.21	8S.48	7	13
7-9	7S.51	9L.77	4	7
6-7	6L.10	7L.60	6	13
3-6	3L.72	6L.75	4	11
3-9	3S.88	9L.82	4	11
1-5	1S.32	5L.31	4	4
3-6	3L.16	6L.32	3	7
2-3	2L.10	3S.31	4	9
4-6	4L.29	6L.25	3	8
1-6	1S.29	6L.48	3	10
2-9	2S.28	9L.27	3	8
5-8	5L.81	8L.67	2	5
1-8	1ctr.	8ctr.	1	2
4-10	4L.56	10S.48	4	4
2-4	2L.35	4S.34	3	5
2-4	2S.28	4L.83	5	7
4-8	4L.83	8S.49	5	7
6-8	6ctr.	8L.46	10	16
1-5	1S.88	5S.62	4	10
2-5	2ctr.	5S.84	5	9
9-10	9S.31	10L.53	7	15
2-5	2S.73	5S.61	2	5
3-6	3S.62	6L.08	3	5
2-4	2S.50	4L.37	2	8
2-6	2S.36	6L.18	8	16

TABLE 3.--Continued.

Symbol	Translocation	Mean position of breaks		Source of data--	
		Ratio	Ratio	Plants, No.	Drawings, No.
061-4-------------------	5-7	5S.54	7L.30	6	14
062-3-------------------	2-5	2L.45	5S.34	2	4
062-11------------------	2-9	2L.21	9S.53	2	5
062-15------------------	2-8	2L.70	8L.26	5	13
062-16------------------	2-5	2S.56	5L.04	1	3
062-16------------------	7-8	7L.15	8L.17	4	10
062-18------------------	5-7	5ctr.	7ctr.	9	16
063-1-------------------	2-10	2L.19	10L.71	7	3
063-1-------------------	3-10	3S.25	10L.71	7	3
064-11------------------	3-9	3L.17	9L.81	12	22
064-13------------------	1-8	1ctr.	8ctr.	2	7
064-18------------------	5-7	5S.61	7S.49	5	10
064-2-------------------	2-3	2S.53	3L.91	2	5
064-20------------------	1-4	1S.23	4L.19	4	8
067-6-------------------	6-9	6S.39	9L.47	8	19
068-12------------------	5-9	5S.71	9L.86	6	13
068-14------------------	1-10	1L.16	10L.79	5	11
070-1-------------------	1-6	1L.40	6L.58	2	6
070-12------------------	1-5	1L.39	5S.71	6	16
071-1-------------------	7-9	7S.70	9L.07	7	15
073-6-------------------	5-10	5L.13	10S.41	2	4
073-8-------------------	4-10	4L.41	10S.74	2	3

TABLE 4.--The breakage points for translocations in series 1 to series 4, arranged in order of the chromosomes and in order along the chromosome arms from left to right.

Trans-location	Symbol	Breakage points Ratio	Ratio	Trans-location	Symbol	Breakage points Ratio	Ratio
1-2--------	7309	1S.90	2S.89	1-3-------	5476	1L.66	3L.87
1-2--------	d	1S.78	2L.56	1-3-------	i	1L.68	3S.30
1-2--------	c	1S.77	2L.33	1-3-------	5267	1L.72	3L.73
1-2--------	e	1S.61	2L.47	1-3-------	4314	1L.81	3L.89
1-2--------	4464	1S.53	2L.28	1-3-------	5242	1L.90	3L.65
1-2--------	b	1S.43	2S.36				
1-2--------	036-7	1S.37	2L.33	1-4-------	h	1S.94	4L.52
1-2--------	5255	1S.25	2S.31	1-4-------	5688	1S.87	4L.45
1-2--------	5896	1S.22	2S.30	1-4-------	002-19	1S.87	4L.42
1-2--------	6891	1S.14	2S.83	1-4-------	4308	1S.65	4L.58
1-2--------	004-11	1S.13	2S.36	1-4-------	b	1S.55	4L.83
1-2--------	8628	1ctr.	2L.49	1-4-------	{ 8598 8602 }	1S.41	4L.81
1-2--------	5946	1ctr.	2ctr.				
1-2--------	028-17	1ctr.	2ctr.	1-4-------	064-20	1S.23	4L.19
1-2--------	4937	1L.10	2S.15	1-4-------	5566	1S.21	4L.26
1-2--------	5453	1L.11	2S.58	1-4-------	8368	1S.14	4S.30
1-2--------	051-1	1L.16	2S.30	1-4-------	K-40	1S.13	4S.42
1-2--------	018-18	1L.16	2L.44	1-4-------	8663		
1-2--------	(9-10a)	1L.21	2L.37	1-4-------	8664	} 1S.09	4S.36
1-2--------	5539	1L.21	2L.61	1-4-------	5910	1S.03	4S.04
1-2--------	015-18	1L.26	2L.26	1-4-------	007-1	1L.04	4L.06
1-2--------	5523	1L.27	2S.62	1-4-------	5629	1L.10	4L.10
1-2--------	041-9	1L.27	2S.57	1-4-------	039-15	1L.14	4S.26
1-2--------	6892	1L.30	2L.35	1-4-------	6422	1L.16	4S.11
1-2--------	6427	1L.43	2L.50	1-4-------	5373	1L.17	4S.29
1-2--------	7211	1L.57	2L.79	1-4-------	f	1L.25	4L.16
1-2--------	6883	1L.63	2L.52	1-4-------	8249	1L.26	4L.63
1-2--------	017-3	1L.67	2S.58	1-4-------	c	1L.33	4S.23
1-2--------	5376	1L.77	2L.08	1-4-------	8563	1L.39	4S.21
				1-4-------	8592	1L.43	4L.69
1-3--------	5883	1S.88	3S.60	1-4-------	e	1L.45	4S.27
1-3--------	5597	1S.77	3L.48	1-4-------	4692	1L.46	4L.15
1-3--------	5982	1S.77	3L.66	1-4-------	015-15	1L.49	4S.40
1-3--------	021-7	1S.72	3L.41	1-4-------	a	1L.51	4S.69
1-3--------	8995	1S.49	3L.06	1-4-------	5756	1L.89	4L.81
1-3--------	k	1S.17	3L.34	1-4-------	5438	1L.93	4L.81
1-3--------	a	1S.15	3L.17	1-4-------	g	1L.95	4L.35
1-3--------	c	1S.14	3L.14				
1-3--------	h	1S.06	3L.04	1-5-------	5045	1S.94	5L.50
1-3--------	013-9	1ctr.	3ctr.	1-5-------	058-2	1S.88	5S.62
1-3--------	6861	1L.04	3L.65	1-5-------	8347	1S.84	5L.51
1-3--------	8048	1L.11	3S.18	1-5-------	4613	1S.78	5L.22
1-3--------	j	1L.11	3L.13	1-5-------	5525	1S.75	5S.53
1-3--------	f	1L.17	3L.11	1-5-------	004-14	1S.75	5L.33
1-3--------	g	1L.17	3L.13	1-5-------	i	1S.71	5S.74
1-3--------	6884	1L.17	3L.19	1-5-------	8972	1S.56	5S.29
1-3--------	024-14	1L.27	3S.49	1-5-------	018-5	1S.53	5L.52
1-3--------	8637	1L.37	3S.50	1-5-------	6899	1S.32	5S.20
1-3--------	4759	1L.39	3L.20	1-5-------	055-4	1S.32	5L.31
1-3--------	e	1L.58	3L.45	1-5-------	4832	1S.20	5L.12
1-3--------	8405	1L.60	1L.31	1-5-------	b	1S.17	5L.10
1-3--------	d	1L.61	3S.75	1-5-------	040-3	1S.17	5L.61

17

TABLE 4.--Continued.

Translocation	Symbol	Breakage points Ratio	Ratio	Translocation	Symbol	Breakage points Ratio	Ratio
1-5--------	5537	1S.11	5S.15	1-7-------	4742	1S.95	7L.03
1-5--------	043-15	1S.10	5L.63	1-7-------	g	1S.79	7S.22
1-5--------	024-5	1S.09	5L.98	1-7-------	4837	1S.73	7L.55
1-5--------	5512	1S.08	5L.70	1-7-------	f	1S.72	7L.80
1-5--------	6197	1S.02	5L.02	1-7-------	4444	1S.65	7S.50
1-5--------	8782	1ctr.	5ctr.	1-7-------	4405	1S.43	7S.46
1-5--------	4331	1L.03	5S.02	1-7-------	6796	1S.40	7S.39
1-5--------	e	1L.03	5L.09	1-7-------	010-12	1S.35	7L.57
1-5--------	6178	1L.04	5L.05	1-7-------	i	1S.31	7L.26
1-5--------	044-10	1L.05	5S.83	1-7-------	021-7	1S.22	7S.56
1-5--------	f	1L.07	5L.09	1-7-------	4302-31	1S.15	7L.12
1-5--------	6401	1L.14	5S.20	1-7-------	f	1S.10	7S.50
1-5--------	7219	1L.15	5S.19	1-7-------	A-37	1L.10	7L.56
1-5--------	h	1L.18	5L.53	1-7-------	5871	1L.12	7L.24
1-5--------	48-34-2	1L.19	5L.76	1-7-------	48-83-15	1L.12	7L.25
1-5--------	5813	1L.21	5S.96	1-7-------	4891	1L.12	7L.69
1-5--------	022-11	1L.27	5L.56	1-7-------	j	1L.20	7L.61
1-5--------	8388	1L.30	5S.25	1-7-------	5339	1L.24	7L.14
1-5--------	c	1L.34	5L.29	1-7-------	a	1L.28	7L.13
1-5--------	070-12	1L.39	5S.71	1-7-------	e	1L.39	7L.11
1-5--------	7212	1L.44	5S.28	1-7-------	c	1L.39	7L.14
1-5--------	4597	1L.52	5S.43	1-7-------	h	1L.46	7L.19
1-5--------	a	1L.52	5S.42	1-7-------	4420	1L.47	7L.90
1-5--------	g	1L.58	5S.85	1-7-------	b	1L.53	7S.12
1-5--------	8041	1L.80	5L.15	1-7-------	d	1L.81	7S.44
1-5--------	7267	1L.92	5L.82	1-7-------	017-10	1L.82	7L.12
				1-7-------	012-10	1L.91	7L.90
1-6--------	8452	.1S.80	6L.52	1-7-------	5693	1L.92	7L.18
1-6--------	8605	} 1S.79	6L.59				
1-6--------	8609			1-8-------	8919	1S.53	8L.44
1-6--------	028-13	1S.56	6L.54	1-8-------	4307-4	1S.42	8L.61
1-6--------	7097	1S.46	6L.62	1-8-------	001-13	1S.39	8L.67
1-6--------	7352	1S.40	6L.60	1-8-------	4685	1S.20	8L.21
1-6--------	e	1S.37	6L.21	1-8-------	6591	1S.18	8S.43
1-6--------	5537	1S.31	6L.22	1-8-------	008-17	1S.16	8L.20
1-6--------	055-10	1S.29	6L.48	1-8-------	5588	1S.10	8S.32
1-6--------	5013	1S.26	6L.28	1-8-------	055-23	1ctr.	8ctr.
1-6--------	5495	1S.25	6S.80	1-8-------	064-13	1ctr.	8ctr.
1-6--------	c	1S.25	6L.27	1-8-------	4676	1L.04	8S.06
1-6--------	6189	1S.23	6L.17	1-8-------	5619	1L.07	8L.16
1-6--------	4986	1S.21	6S.78	1-8-------	5634	1L.08	8S.28
1-6--------	5077	1S.20	6L.60	1-8-------	5384	1L.10	8L.59
1-6--------	5072	1S.17	6L.45	1-8-------	8683	1L.11	8ctr.
1-6--------	h	1L.03	6L.17	1-8-------	8640	1L.11	8L.16
1-6--------	d	1L.13	6S.74	1-8-------	020-19	1L.11	8L.38
1-6--------	g	1L.16	6L.84	1-8-------	4748	1L.12	8L.15
1-6--------	a	1L.20	6L.54	1-8-------	036-4	1L.18	8L.59
1-6--------	8415	1L.29	6S.82	1-8-------	005-7	1L.22	8L.78
1-6--------	f	1L.32	6L.42	1-8-------	7509	1L.28	8L.21
1-6--------	070-1	1L.40	6L.58	1-8-------	5752	1L.36	8L.25
1-6--------	5225	1L.61	6L.72	1-8-------	a	1L.41	8S.52
1-6--------	4456	1L.71	6L.30	1-8-------	026-2	1L.49	8L.80
1-6--------	8650	} 1L.79	6L.91	1-8-------	6766	1L.54	8L.77
1-6--------	8658			1-8-------	b	1L.59	8L.82

TABLE 4.--Continued.

Trans-location	Symbol	Breakage points Ratio	Ratio	Trans-location	Symbol	Breakage points Ratio	Ratio
1-8--------	5821	1L.65	8L.31	2-1-------	5523	2S.62	1L.27
1-8--------	6697	1L.89	8L.52	2-1-------	5453	2S.58	1L.11
1-8--------	5910	1L.93	8L.67	2-1-------	017-3	2S.58	1L.67
1-8--------	c	1L.94	8L.89	2-1-------	041-9	2S.57	1L.27
1-8--------	5704	1L.96	8S.67	2-1-------	b	2S.36	1S.43
				2-1-------	004-11	2S.36	1S.13
1-9--------	024-7	1S.71	9L.13	2-1-------	5255	2S.31	1S.25
1-9--------	8265	1S.58	9L.27	2-1-------	5896	2S.30	1S.22
1-9--------	8302	1S.55	9L.29	2-1-------	051-1	2S.30	1L.16
1-9--------	8129	1S.53	9L.27	2-1-------	4937	2S.15	1L.10
1-9--------	48-34-2	1S.51	9L.24	2-1-------	5946	2ctr.	1ctr.
1-9--------	8001	1S.51	9L.24	2-1-------	028-17	2ctr.	1ctr.
1-9--------	c	1S.48	9L.22	2-1-------	5376	2L.08	1L.77
1-9--------	7535	1S.33	9S.27	2-1-------	015-18	2L.26	1L.26
1-9--------	8918	1S.21	9L.20	2-1-------	4464	2L.28	1S.53
1-9--------	6762	1S.16	9L.53	2-1-------	c	2L.33	1S.77
1-9--------	a	1S.13	9L.15	2-1-------	036-7	2L.33	1S.37
1-9--------	8460	1S.13	9L.24	2-1-------	6892	2L.35	1L.30
1-9--------	8824	1S.06	9L.83	2-1-------	(9-10a)	2L.37	1L.21
1-9--------	5622	1L.10	9L.12	2-1-------	018-18	2L.44	1L.16
1-9--------	4995	1L.19	9S.20	2-1-------	e	2L.47	1S.61
1-9--------	9021	1L.26	9L.83	2-1-------	8628	2L.49	1ctr.
1-9--------	8886	}1L.33	9L.23	2-1-------	6427	2L.50	1L.43
1-9--------	8889			2-1-------	6883	2L.52	1L.63
1-9--------	4997	1L.37	9S.28	2-1-------	d	2L.56	1S.78
1-9--------	d	1L.42	9L.25	2-1-------	5539	2L.61	1L.21
1-9--------	b	1L.50	9L.60	2-1-------	7211	2L.79	1L.57
1-9--------	4398	1L.51	9S.19				
1-9--------	8389	1L.74	9L.13	2-3-------	023-5	2S.80	3L.70
1-9--------	6328	1L.79	9L.40	2-3-------	8662	2S.78	3L.83
1-9--------	035-10	1L.89	9S.67	2-3-------	e	2S.76	3L.48
				2-3-------	5800	2S.73	3S.81
1-10-------	g	1S.80	10L.21	2-3-------	5304	2S.62	3L.29
1-10-------	007-19	1S.07	10L.08	2-3-------	064-2	2S.53	3L.91
1-10-------	f	1S.04	10L.30	2-3-------	c	2S.46	3S.52
1-10-------	4885	1ctr.	10ctr.	2-3-------	6270	2S.46	3L.60
1-10-------	8770	1L.09	10L.38	2-3-------	014-12	2S.43	3L.51
1-10-------	e	1L.16	10L.31	2-3-------	6862	2S.39	3L.20
1-10-------	068-14	1L.16	10L.79	2-3-------	4369	2S.19	3S.26
1-10-------	5273	1L.17	10L.69	2-3-------	010-10	2S.17	3L.13
1-10-------	b	1L.19	10S.39	2-3-------	8857	2S.08	3L.24
1-10-------	a	1L.29	10L.33	2-3-------	4301-111	2ctr.	3ctr.
1-10-------	c	1L.43	10L.74	2-3-------	023-2	2ctr.	3ctr.
1-10-------	8465	}1L.45	10L.76	2-3-------	055-7	2L.10	3S.31
1-10-------	8491			2-3-------	005-14	2L.12	3S.29
1-10-------	d	1L.50	10L.68	2-3-------	h	2L.14	3L.07
1-10-------	015-9	1L.67	10S.46	2-3-------	8483	2L.14	3L.12
1-10-------	8375	1L.69	10L.64	2-3-------	I-10	2L.19	3S.51
1-10-------	001-3	1L.86	10L.48	2-3-------	g	2L.21	3S.21
1-10-------	13-29	1L.93	10L.26	2-3-------	8553	}2L.26	3L.23
				2-3-------	8556		
2-1--------	7309	2S.89	1S.90	2-3-------	7285	2L.26	3L.39
2-1--------	6891	2S.83	1S.14	2-3-------	033-4	2L.27	3L.23

19

TABLE 4.--Continued.

Trans-location	Symbol	Breakage points Ratio	Ratio	Trans-location	Symbol	Breakage points Ratio	Ratio
2-3	f	2L.35	3S.60	2-5	010-17	2S.19	5L.21
2-3	b	2L.45	3L.08	2-5	059-1	2ctr.	5S.84
2-3	8628	2L.54	3L.44	2-5	b	2L.06	5S.09
2-3	d	2L.67	3L.48	2-5	6580	2L.09	5S.09
2-3	4303-74	2L.73	3L.68	2-5	5098	2L.13	5S.23
2-3	6750	2L.76	3S.53	2-5	a	2L.14	5L.15
2-3	011-6	2L.77	3L.48	2-5	c	2L.16	5S.48
2-3	(3-6a)	2L.79	3S.90	2-5	015-3	2L.16	5S.69
2-3	6284	2L.81	3L.75	2-5	002-16	2L.25	5L.35
				2-5	023-15	2L.28	5S.30
2-4	5157	2S.86	4L.07	2-5	A-16	2L.34	5S.20
2-4	4413	2S.84	4S.32	2-5	032-9	2L.40	5S.31
2-4	8682	2S.70	4S.25	2-5	062-3	2L.45	5S.34
2-4	6994	2S.59	4L.95	2-5	5876	2L.47	5L.46
2-4	8865	2S.52	4L.27	2-5	5645	2L.60	5S.85
2-4	060-8	2S.50	4L.37	2-5	6885	2L.63	5S.79
2-4	018-3	2S.38	4L.47	2-5	5602	2L.73	5L.77
2-4	057-21	2S.28	4L.83	2-5	8321	2L.86	5L.11
2-4	5495	2S.27	4L.10	2-5	f	2L.91	5L.10
2-4	j	2ctr.	4ctr.	2-5	d	2L.91	5L.86
2-4	8407	2ctr.	4ctr.	2-5	4578	2L.92	5S.71
2-4	052-13	2ctr.	4ctr.				
2-4	g	2L.13	4S.31	2-6	4394	2S.91	6L.12
2-4	010-4	2L.13	4S.31	2-6	8786	2S.90	6S.77
2-4	k	2L.13	4L.04	2-6	001-15	2S.72	6S.87
2-4	004-13	2L.14	4S.51	2-6	b	2S.69	6L.49
2-4	4374	2L.15	4L.23	2-6	008-5	2S.40	6L.26
2-4	8027	2L.15	4L.43	2-6	4372	2S.37	6S.81
2-4	d	2L.17	4L.45	2-6	060-5	2S.36	6L.18
2-4	5951	2L.18	4S.26	2-6	027-4	2S.34	6L.11
2-4	a	2L.30	4L.21	2-6	5472 / 5473	} 2S.25	6L.15
2-4	e	2L.31	4S.47	2-6	6671	2S.22	6L.22
2-4	m	2L.34	4S.47	2-6	e	2L.18	6L.20
2-4	057-19	2L.35	4S.51	2-6	6931	2L.24	6L.23
2-4	017-18	2L.39	4L.19	2-6	"78"	2L.24	6L.29
2-4	6266	2L.40	4L.27	2-6	5648	2L.25	6L.19
2-4	011-7	2L.53	4L.76	2-6	a	2L.28	6L.20
2-4	b	2L.59	4S.40	2-6	c	2L.37	6L.25
2-4	f	2L.75	4L.12	2-6	d	2L.41	6L.45
2-4	052-15	2L.75	4L.66	2-6	9002 / 9009	} 2L.57	6L.50
2-4	016-19	2L.81	4S.79	2-6	4717	2L.77	6L.27
2-4	014-11	2L.81	4S.16	2-6	f	2L.79	6L.87
2-4	c	2L.81	4S.09	2-6	014-11	2L.81	6L.20
2-4	b	2L.81	4L.53	2-6	5419	2L.82	6S.79
				2-6	8441	2L.94	6S.79
2-5	g	2S.79	5S.24				
2-5	059-17	2S.73	5S.61	2-7	5279	2S.93	7L.25
2-5	019-1	2S.67	5S.51	2-7	5144	2S.35	7L.08
2-5	062-16	2S.56	5L.04	2-7	022-4	2S.30	7L.24
2-5	4741	2S.47	5L.47	2-7	8045	2S.12	7L.06
2-5	009-19	2S.29	5L.11	2-7	d	2L.16	7L.18
2-5	025-4	2S.26	5L.78				
2-5	e	2S.19	5S.28				

TABLE 4.--Continued.

Trans-location	Symbol	Breakage points		Trans-location	Symbol	Breakage points	
		Ratio	Ratio			Ratio	Ratio
2-7--------	48-74-1	2L.22	7S.47	2-10------	043-10	2S.89	10L.40
2-7--------	4400	2L.24	7L.32	2-10------	5651	2S.71	10L.62
2-7--------	8465	2L.27	7S.56	2-10------	b	2S.50	10L.75
2-7--------	f	2L.30	7L.68	2-10------	8864	2S.10	10L.76
2-7--------	b	2L.37	7L.12	2-10------	4484	2S.09	10L.14
2-7--------	c	2L.47	7S.34	2-10------	5830	2L.12	10L.12
2-7--------	4519	2L.65	7L.66	2-10------	a	2L.16	10L.55
2-7--------	5783	2L.66	7L.10	2-10------	063-1	2L.19	10L.71
2-7--------	038-12	2L.75	7S.68	2-10------	I-3	2L.30	10S.40
2-7--------	8322	2L.76	7L.74	2-10------	5561	2L.35	10S.16
2-7--------	e	2L.82	7L.63	2-10------	011-9	2L.39	10L.76
					8219	2L.50	10L.35
2-8--------	013-17	2S.89	8L.61	2-10------	6853	2L.79	10L.86
2-8--------	4711	2S.86	8L.67	2-10------	035-2	2L.85	10L.49
2-8--------	011-20	2S.58	8L.28				
2-8--------	8458	2S.22	8L.32	3-1-------	d	3S.75	1L.61
2-8--------	c	2S.15	8S.11	3-1-------	5883	3S.60	1S.88
2-8--------	f	2ctr.	8ctr.	3-1-------	8637	3S.50	1L.37
2-8--------	006-10	2ctr.	8ctr.	3-1-------	024-14	3S.49	1L.27
2-8--------	d	2L.05	8L.10	3-1-------	i	3S.30	1L.68
2-8--------	e	2L.07	8L.10	3-1-------	8048	3S.18	1L.11
2-8--------	4414	2L.12	8L.14	3-1-------	013-9	3ctr.	1ctr.
2-8--------	7069	2L.13	8L.14	3-1-------	h	3L.04	1S.06
2-8--------	8428	2L.16	8L.10	3-1-------	8995	3L.06	1S.49
2-8--------	003-5	2L.19	8S.72	3-1-------	f	3L.11	1L.17
2-8--------	b	2L.20	8L.18	3-1-------	j	3L.13	1L.11
2-8--------	5454	2L.21	8S.39	3-1-------	g	3L.13	1L.17
2-8--------	h	2L 23	8L.22	3-1-------	c	3L.14	1S.14
2-8--------	5484	2L.24	8S.58	3-1-------	a	3L.17	1S.15
2-8--------	031-7	2L.30	8S.44	3-1-------	6884	3L.19	1L.17
2-8--------	"84"	2L.32	8L.30	3-1-------	4759	3L.20	1L.39
2-8--------	051-15	2L.62	8L.48	3-1-------	8405	3L.31	1L.60
2-8--------	062-15	2L.70	8L.26	3-1-------	k	3L.34	1S.17
2-8--------	g	2L.71	8S.71	3-1-------	021-7	3L.41	1S.72
2-8--------	051-7	2L.83	8L.74	3-1-------	e	3L.45	1L.58
2-8--------	48-45-6	2L.84	8L.68	3-1-------	5597	3L.48	1S.77
2-8--------	8376	2L.95	8L.03	3-1-------	6861	3L.65	1L.04
2-8--------	037-5	2L.95	8L.54	3-1-------	5242	3L.65	1L.90
				3-1-------	5982	3L.66	1S.77
2-9--------	7096	2S.57	9L.66	3-1-------	5267	3L.73	1L.72
2-9--------	c	2S.49	9S.33	3-1-------	5476	3L.87	1L.66
2-9--------	a	2S.36	9L.58	3-1-------	4314	3L.89	1L.81
2-9--------	055-14	2S.28	9L.27				
2-9--------	(2-8a)	2S.27	9L.72	3-2-------	(3-6a)	3S.90	2L.79
2-9--------	5711	2S.24	9L.23	3-2-------	5800	3S.81	2S.73
2-9--------	b	2S.18	9L.22	3-2-------	f	3S.60	2L.35
2-9--------	062-11	2L.21	9S.53	3-2-------	6750	3S.53	2L.76
2-9--------	5257	2L.28	9L.20	3-2-------	c	3S.52	2S.46
2-9--------	6656	2L.32	9S.31	3-2-------	I-10	3S.51	2L.19
2-9--------	5831	2L.63	9S.72	3-2-------	055-7	3S.31	2L.10
2-9--------	5208	2L.76	9L.68	3-2-------	005-14	3S.29	2L.12
2-9--------	d	2L.83	9L.27	3-2-------	4369	3S.26	2S.19

21

TABLE 4.--Continued.

Trans-location	Symbol	Breakage points Ratio	Breakage points Ratio	Trans-location	Symbol	Breakage points Ratio	Breakage points Ratio
3-2	g	3S.21	2L.21	3-5	039-13	3L.13	5L.14
3-2	4301-111	3ctr.	2ctr.	3-5	5874	3L.16	5L.21
3-2	023-2	3ctr.	2ctr.	3-5	5521	3L.17	5L.48
3-2	h	3L.07	2L.14	3-5	015-18	3L.18	5S.21
3-2	b	3L.08	2L.45	3-5	13-104	3L.23	5L.20
3-2	8483	3L.12	2L.14	3-5	a	3L.28	5L.60
3-2	010-10	3L.13	2S.17	3-5	h	3L.55	5L.22
3-2	6862	3L.20	2S.39	3-5	b	3L.61	5L.57
3-2	8553 }			3-5	c	3L.62	5L.27
3-2	8556 }	3L.23	2L.26	3-5	7043	3L.63	5L.61
3-2	033-4	3L.23	2L.27	3-5	8351	3L.75	5L.68
3-2	8857	3L.24	2S.08	3-5	6346	3L.94	5L.83
3-2	5304	3L.29	2S.62				
3-2	7285	3L.39	2L.26	3-6	b	3S.73	6S.82
3-2	8628	3L.44	2L.54	3-6	060-4	3S.62	6L.08
3-2	d	3L.48	2L.67	3-6	4349	3S.58	6L.70
3-2	e	3L.48	2S.76	3-6	c	3S.56	6L.54
3-2	011-6	3L.48	2L.77	3-6	016-17	3S.48	6L.30
3-2	014-12	3L.51	2S.43	3-6	032-3	3S.41	6S.78
3-2	6270	3L.60	2S.46	3-6	030-8	3S.27	6S.81
3-2	4303-74	3L.68	2L.73	3-6	8963	3S.23	6L.14
3-2	023-5	3L.70	2S.80	3-6	a	3L.06	6L.30
3-2	6284	3L.75	2L.81	3-6	7067	3L.07	6L.75
3-2	8662	3L.83	2S.78	3-6	6349	3L.10	6L.15
3-2	064-2	3L.91	2S.53	3-6	055-5	3L.16	6L.32
				3-6	8145	3L.17	6L.26
3-4	8969	3S.75	4L.75	3-6	5368	3L.22	6L.20
3-4	8397	3S.74	4S.55	3-6	d	3L.23	6L.82
3-4	8634	3S.71	4L.75	3-6	5201	3L.26	6L.21
3-4	5156	3S.47	4L.67	3-6	003-2	3L.35	6L.15
3-4	5920	3S.28	4L.73	3-6	6566	3L.41	6L.35
3-4	012-16	3S.27	4S.30	3-6	8672	3L.47	6L.87
3-4	4662	3S.24	4S.67	3-6	7162	3L.52	6L.53
3-4	4726	3S.16	4L.15	3-6	054-12	3L.72	6L.75
3-4	5891	3ctr.	4ctr.	3-6	6266	3L.85	6S.80
3-4	A-21	3L.07	4L.85	3-7	b	3S.92	7L.03
3-4	006-17	3L.10	4S.45	3-7	001-15	3S.38	7L.30
3-4	037-9	3L.10	4L.14	3-7	004-7	3S.38	7L.26
3-4	8443	3L.12	4L.13	3-7	a	3S.25	7L.18
3-4	4713	3L.21	4S.59	3-7	4670	3S.20	7L.76
3-4	6534	3L.48	4L.89	3-7	4773	3S.11	7L.07
				3-7	5724	3ctr.	7ctr.
3-5	4635	3S.44	5S.48	3-7	8541	3ctr.	7ctr.
3-5	e	3S.34	5S.16	3-7	6557	3L.06	7S.59
3-5	6473	3S.32	5S.26	3-7	5955	3L.10	7L.58
3-5	6462	3S.31	5L.47	3-7	029-3	3L.11	7L.13
3-5	4873	3S.24	5S.16	3-7	5378	3L.13	7L.73
3-5	4880	3ctr.	5ctr.	3-7	e	3L.25	7S.56
3-5	4898	3ctr.	5ctr.	3-7	6466	3L.36	7L.14
3-5	6695	3ctr.	5ctr.	3-7	c	3L.46	7L.45
3-5	g	3L.01	5S.73	3-7	5471	3L.64	7L.58
3-5	8104	3L.05	5L.08	3-7	d	3L.64	7L.81
3-5	8528	3L.06	5L.72	3-7	8006	3L.88	7L.90

TABLE 4.--Continued.

Trans-location	Symbol	Breakage points		Trans-location	Symbol	Breakage points	
		Ratio	Ratio			Ratio	Ratio
3-8--------	024-11	3S.65	8L.49	3-9------	4727	3L.54	9L.42
3-8--------	6373	3S.53	8L.68	3-9------	f	3L.63	9S.69
3-8--------	e	3S.36	8L.21	3-9------	8562		
3-8--------	4626	3S.30	8L.31	3-9------	8568		
3-8--------	6439	3S.30	8L.15	3-9------	8572	} 3L.65	9L.22
3-8--------	8666			3-9------	8574		
3-8--------	8667	} 3S.30	8L.14	3-9------	4963	3L.76	9L.57
3-8--------	8670						
3-8--------	8367	3S.28	8S.52	3-10-----	7464	3S.49	10L.60
3-8--------	5558	3S.26	8S.74	3-10-----	8412	3S.39	10S.36
3-8--------	c	3S.23	8L.85	3-10-----	8349	3S.38	10ctr.
3-8--------	015-17	3S.12	8S.13	3-10-----	4382	3S.38	10L.29
3-8--------	5830	3S.06	8L.13	3-10-----	063-1	3S.25	10L.71
3-8--------	4303-12	3ctr.	8ctr.	3-10-----	4383	3S.23	10L.30
3-8--------	4872	3ctr.	8ctr.	3-10-----	5892	3S.17	10L.25
3-8--------	043-14	3L.02	8S.40	3-10-----	a	3L.16	10L.22
3-8--------	5583	3L.06	8L.05	3-10-----	b	3L.19	10L.27
3-8--------	7362	3L.07	8L.69	3-10-----	c	3L.22	10L.30
3-8--------	f	3L.08	8L.10	3-10-----	6691	3L.30	10L.87
3-8--------	g	3L.12	8L.19	3-10-----	036-15	3L.48	10L.64
3-8--------	007-8	3L.14	8L.11	3-10-----	044-10	3L.77	10L.72
3-8--------	b	3L.16	8L.23				
3-8--------	8023	3L.18	8L.16	4-1------	a	4S.69	1L.51
3-8--------	4874	3L.28	8L.32	4-1------	K-40	4S.42	1S.13
3-8--------	012-10	3L.39	8L.23	4-1------	015-15	4S.40	1L.49
3-8--------	a	3L.41	8L.61	4-1------	8663	} 4S.36	1S.09
3-8--------	6261	3L.49	8L.40	4-1------	8664		
3-8--------	h	3L.53	8S.46	4-1------	8368	4S.30	1S.14
3-8--------	B-31	3L.63	8S.06	4-1------	5373	4S.29	1L.17
3-8--------	8350	3L.75	8S.60	4-1------	e	4S.27	1L.45
3-8--------	4340	3L.88	8L.72	4-1------	039-15	4S.26	1L.14
3-8--------	4301-39	3L.92	8L.82	4-1------	c	4S.23	1L.33
				4-1------	5863	4S.21	1L.39
3-9--------	054-18	3S.88	9L.82	4-1------	6422	4S.11	1L.16
3-9--------	6722	3S.66	9S.66	4-1------	5910	4S.04	1S.03
3-9--------	7041	3S.59	9L.70	4-1------	007-1	4L.06	1L.04
3-9--------	5643	3S.55	9L.64	4-1------	5629	4L.10	1L.10
3-9--------	8447	3S.44	9L.14	4-1------	4692	4L.15	1L.46
3-9--------	030-2	3S.39	9L.30	4-1------	f	4L.16	1L.25
3-9--------	8465	3S.27	9L.41	4-1------	064-20	4L.19	1S.23
3-9--------	8032	3S.26	9L.96	4-1------	5566	4L.26	1S.21
3-9--------	020-5	3ctr.	9ctr.	4-1------	g	4L.35	1L.95
3-9--------	5775	3L.09	9S.24	4-1------	002-19	4L.42	1S.87
3-9--------	c	3L.09	9L.12	4-1------	5680	4L.45	1S.87
3-9--------	h	3L.09	9L.33	4-1------	h	4L.52	1S.94
3-9--------	a	3L.11	9L.16	4-1------	4308	4L.58	1S.65
3-9--------	d	3L.13	9L.26	4-1------	8249	4L.63	1L.26
3-9--------	064-11	3L.17	9L.81	4-1------	8592	4L.69	1L.43
3-9--------	g	3L.40	9L.14	4-1------	8598	} 4L.81	1S.41
3-9--------	034-11	3L.46	9S.36	4-1------	8602		
3-9--------	b	3L.48	9L.53	4-1------	5756	4L.81	1L.89
3-9--------	5285	3L.51	9L.49	4-1------	5438	4L.81	1L.93
				4-1------	b	4L.83	1S.55

23

TABLE 4.--Continued.

Trans-location	Symbol	Breakage points Ratio	Ratio	Trans-location	Symbol	Breakage points Ratio	Ratio
4-2--------	016-19	4S.79	2L.81	4-5------	8108	4S.37	5S.72
4-2--------	004-13	4S.51	2L.14	4-5------	8006	4S.37	5S.18
4-2--------	057-19	4S.51	2L.35	4-5------	5529	4S.37	5L.46
4-2--------	e	4S.47	2L.31	4-5------	8069	4S.34	5S.71
4-2--------	m	4S.47	2L.34	4-5------	c	4S.34	5L.27
4-2--------	l	4S.40	2L.59	4-5------	6831	4S.32	5S.59
4-2--------	4413	4S.32	2S.84	4-5------	6560	4S.32	5S.21
4-2--------	g	4S.31	2L.13	4-5------	002-12	4S.29	5S.36
4-2--------	010-4	4S.31	2L.13	4-5------	4305-8	4S.27	5L.28
4-2--------	5951	4S.26	2L.18	4-5------	4472	4S.25	5S.19
4-2--------	8682	4S.25	2S.70	4-5------	d	4S.21	5L.22
4-2--------	014-11	4S.16	2L.81	4-5------	k	4S.06	5L.13
4-2--------	c	4S.09	2L.81	4-5------	8891	4ctr.	5ctr.
4-2--------	j	4ctr.	2ctr.	4-5------	7078	4L.05	5L.10
4-2--------	8407	4ctr.	2ctr.	4-5------	i	4L.10	5S.15
4-2--------	052-13	4ctr.	2ctr.	4-5------	h	4L.13	5L.08
4-2--------	k	4L.04	2L.13	4-5------	8955	4L.17	5S.32
4-2--------	5157	4L.07	2S.86	4-5------	a	4L.19	5S.29
4-2--------	5495	4L.10	2S.27	4-5------	j	4L.21	5L.36
4-2--------	f	4L.12	2L.75	4-5------	007-18	4L.22	5L.39
4-2--------	017-18	4L.19	2L.39	4-5------	g	4L.27	5S.70
4-2--------	a	4L.21	2L.30	4-5------	8622	4L.30	5L.52
4-2--------	4374	4L.23	2L.15	4-5------	044-8	4L.38	5L.33
4-2--------	8865	4L.27	2S.52	4-5------	006-7	4L.43	5S.25
4-2--------	6266	4L.27	2L.40	4-5------	7136	4L.45	5L.33
4-2--------	060-8	4L.37	2S.50	4-5------	6743	4L.56	5S.59
4-2--------	8027	4L.43	2L.15	4-5------	018-4	4L.61	5L.67
4-2--------	d	4L.45	2L.17	4-5------	027-10	4L.61	5L.79
4-2--------	018-3	4L.47	2S.38	4-5------	021-3	4L.62	5S.71
4-2--------	b	4L.53	2L.81	4-5------	8395	4L.63	5L.82
4-2--------	052-15	4L.66	2L.75	4-5------	b	4L.76	5L.68
4-2--------	011-7	4L.76	2L.53				
4-2--------	057-21	4L.83	2S.28				
4-2--------	6994	4L.95	2S.59	4-6------	4461	4S.86	6L.17
				4-6------	b	4S.80	6L.16
4-3--------	4662	4S.67	3S.24	4-6------	e	4S.62	6L.56
4-3--------	4713	4S.59	3L.21	4-6------	7328	4S.53	6S.89
4-3--------	8397	4S.55	3S.74	4-6------	8380	4S.47	6L.18
4-3--------	006-17	4S.45	3L.10	4-6------	5227	4S.46	6S.84
4-3--------	012-16	4S.30	3S.27	4-6------	025-12	4S.44	6L.34
4-3--------	5891	4ctr.	3ctr.	4-6------	4341	4S.37	6S.81
4-3--------	8443	4L.13	3L.12	4-6------	c	4S.33	6S.83
4-3--------	037-9	4L.14	3L.10	4-6------	011-16	4S.31	6L.33
4-3--------	4726	4L.15	3S.16	4-6------	4447	4S.28	6L.14
4-3--------	5156	4L.67	3S.47	4-6------	8591 }	4L.17	6L.24
4-3--------	5920	4L.73	3S.28	4-6------	8593		
4-3--------	8969	4L.75	3S.75	4-6------	6623	4L.18	6L.31
4-3--------	8634	4L.75	3S.71	4-6------	055-8	4L.29	6L.25
4-3--------	A-21	4L.85	3L.07	4-6------	8428	4L.32	6L.28
4-3--------	6534	4L.89	3L.48	4-6------	8764	4L.32	6L.90
				4-6------	a	4L.37	6L.43
4-5--------	e	4S.41	5S.32	4-6------	d	4L.49	6L.53
4-5--------	g	4S.38	5L.30	4-6------	003-16	4L.50	6S.90

TABLE 4.--Continued.

Trans-location	Symbol	Breakage points Ratio	Ratio	Trans-location	Symbol	Breakage points Ratio	Ratio
4-6-------	7037	4L.61	6S.77	4-9------	5657	4L.33	9S.25
4-6-------	8927	4L.70	6L.18	4-9------	5884	4L.40	9L.49
4-6-------	038-11	4L.78	6L.29	4-9------	4333	4L.41	9L.93
4-6-------	8339	4L.87	6L.79	4-9------	f	4L.55	9L.18
				4-9------	5828	4L.61	9L.27
4-7-------	8103	4S.81	7L.76	4-9------	5788	4L.72	9L.82
4-7-------	6575	4S.38	7S.32	4-9------	5574	4L.80	9L.87
4-7-------	a	4S.32	7L.06	4-9------	c	4L.82	9L.29
4-7-------	48-40-8	4S.32	7L.64	4-9------	b	4L.90	9L.29
4-7-------	7347	4S.31	7L.66	4-9------	8636 }	4L.94	9S.09
4-7-------	7108	4S.17	7S.45	4-9------	8649 }		
4-7-------	4698	4L.08	7L.74				
4-7-------	7067	4L.17	7S.60	4-10-----	c	4S.64	10L.18
4-7-------	027-17	4L.17	7L.31	4-10-----	9028 }	4S.57	10L.89
4-7-------	8374	4L.24	7L.55	4-10-----	9029 }		
4-7-------	052-13	4L.27	7L.25	4-10-----	8541	4S.45	10ctr.
4-7-------	4483	4L.39	7L.61	4-10-----	d	4S.36	10L.36
4-7-------	008-16	4L.86	7L.17	4-10-----	6662	4L.04	10L.03
				4-10-----	e	4L.14	10L.14
4-8-------	036-16	4S.66	8L.69	4-10-----	b	4L.15	10L.60
4-8-------	a	4S.59	8L.19	4-10-----	021-5	4L.34	10L.33
4-8-------	8987	4S.58	8L.76	4-10-----	073-8	4L.41	10S.74
4-8-------	8603 }			4-10-----	6587	4L.55	10L.51
4-8-------	8606 }			4-10-----	057-14	4L.56	10S.48
4-8-------	8607 }	4S.42	8L.35	4-10-----	024-16	4L.75	10L.18
4-8-------	8608 }			4-10-----	f	4L.94	10L.14
4-8-------	4356	4S.40	8L.14				
4-8-------	8004	4S.27	8L.84	5-1------	5813	5S.96	1L.21
4-8-------	8677	4S.24	8L.79	5-1------	g	5S.85	1L.58
4-8-------	5339	4S.22	8L.71	5-1------	044-10	5S.83	1L.05
4-8-------	8456	4S.22	8L.75	5-1------	i	5S.74	1S.71
4-8-------	b	4S.18	8L.16	5-1------	070-12	5S.71	1S.39
4-8-------	8163	4S.17	8L.86	5-1------	058-2	5S.62	1S.88
4-8-------	6063	4S.02	8L.05	5-1------	5525	5S.53	1S.75
4-8-------	004-12	4L.10	8L.10	5-1------	4597	5S.43	1L.52
4-8-------	016-6	4L.42	8S.45	5-1------	a	5S.42	1L.52
4-8-------	5412	4L.59	8L.95	5-1------	8972	5S.29	1S.56
4-8-------	6926	4L.60	8L.71	5-1------	7212	5S.28	1L.44
4-8-------	6363	4L.76	8L.30	5-1------	8388	5S.25	1L.30
4-8-------	057-21	4L.83	8S.49	5-1------	6899	5S.20	1S.32
				5-1------	6401	5S.20	1L.14
4-9-------	e	4S.53	9L.26	5-1------	7219	5S.19	1L.15
4-9-------	4307-12	4S.48	9L.55	5-1------	5537	5S.15	1S.11
4-9-------	g	4S.27	9L.27	5-1------	4331	5S.02	1L.03
4-9-------	5918	4S.24	9L.18	5-1------	8782	5ctr.	1ctr.
4-9-------	4304-82	4S.22	9L.37	5-1------	6197	5L.02	1S.02
4-9-------	6222	4L.03	9S.68	5-1------	6178	5L.05	1L.04
4-9-------	6504	4L.09	9S.83	5-1------	e	5L.09	1L.03
4-9-------	d	4L.12	9L.17	5-1------	f	5L.09	1L.07
4-9-------	a	4L.16	9L.58	5-1------	b	5L.10	1S.17
4-9-------	004-7	4L.28	9L.26	5-1------	4832	5L.12	1S.20
4-9-------	4373	4L.29	9L.39	5-1------	8041	5L.15	1L.80

TABLE 4.--Continued.

Trans-location	Symbol	Breakage points Ratio	Ratio	Trans-location	Symbol	Breakage points Ratio	Ratio
5-1-------	4613	5L.22	1S.78	5-3------	4898	5ctr.	3ctr.
5-1-------	c	5L.29	1L.34	5-3------	6695	5ctr.	3ctr.
5-1-------	055-4	5L.31	1S.32	5-3------	8104	5L.08	3L.05
5-1-------	004-14	5L.33	1S.75	5-3------	039-13	5L.14	3L.13
5-1-------	5045	5L.50	1S.94	5-3------	B-104	5L.20	3L.23
5-1-------	8347	5L.51	1S.84	5-3------	5874	5L.21	3L.16
5-1-------	018-5	5L.52	1S.53	5-3------	h	5L.22	3L.55
5-1-------	h	5L.53	1L.18	5-3------	c	5L.27	3L.62
5-1-------	022-11	5L.56	1L.27	5-3------	6462	5L.47	3S.31
5-1-------	040-3	5L.61	1S.17	5-3------	5521	5L.48	3L.17
5-1-------	043-15	5L.63	1S.10	5-3------	b	5L.57	3L.61
5-1-------	5512	5L.70	1S.08	5-3------	a	5L.60	3L.28
5-1-------	48-34-2	5L.76	1L.19	5-3------	8351	5L.60	3L.75
5-1-------	7267	5L.82	1L.92	5-3------	7043	5L.61	3L.63
5-1-------	024-5	5L.98	1S.05	5-3------	8528	5L.72	3L.06
				5-3------	6346	5L.83	3L.94
5-2-------	5645	5S.85	2L.60				
5-2-------	059-1	5S.84	2ctr.	5-4------	8108	5S.72	4S.37
5-2-------	6885	5S.79	2L.63	5-4------	8069	5S.71	4S.34
5-2-------	4578	5S.71	2L.92	5-4------	021-3	5S.71	4L.62
5-2-------	015-3	5S.69	2L.16	5-4------	g	5S.70	4L.27
5-2-------	059-17	5S.61	2S.73	5-4------	6831	5S.59	4S.32
5-2-------	019-1	5S.51	2S.67	5-4------	6743	5S.59	4L.56
5-2-------	c	5S.48	2L.16	5-4------	002-12	5S.36	4S.29
5-2-------	062-3	5S.34	2L.45	5-4------	e	5S.32	4S.41
5-2-------	032-9	5S.31	2L.40	5-4------	8955	5S.32	4L.17
5-2-------	023-15	5S.30	2L.28	5-4------	a	5S.29	4L.19
5-2-------	e	5S.28	2L.19	5-4------	006-7	5S.25	4L.43
5-2-------	g	5S.24	2S.79	5-4------	6560	5S.21	4S.32
5-2-------	5098	5S.23	2L.13	5-4------	4472	5S.19	4S.25
5-2-------	A-16	5S.20	2L.34	5-4------	8006	5S.18	4S.37
5-2-------	b	5S.09	2L.06	5-4------	i	5S.15	4L.10
5-2-------	6580	5S.09	2L.09	5-4------	8891	5ctr.	4ctr.
5-2-------	062-16	5L.04	2S.56	5-4------	h	5L.08	4L.13
5-2-------	f	5L.10	2L.91	5-4------	7078	5L.10	4L.05
5-2-------	009-19	5L.11	2S.29	5-4------	k	5L.13	4S.06
5-2-------	8321	5L.11	2L.86	5-4------	d	5L.22	4S.21
5-2-------	a	5L.15	2L.14	5-4------	c	5L.27	4S.34
5-2-------	e	5L.21	2S.19	5-4------	4305-8	5L.28	4S.27
5-2-------	002-16	5L.35	2L.25	5-4------	g	5L.30	4S.38
5-2-------	5876	5L.46	2L.47	5-4------	004-8	5L.33	4L.38
5-2-------	4741	5L.47	2S.47	5-4------	7136	5L.33	4L.45
5-2-------	5602	5L.77	2L.73	5-4------	j	5L.36	4L.21
5-2-------	025-4	5L.78	2S.26	5-4------	007-18	5L.39	4L.22
5-2-------	d	5L.86	2L.91	5-4------	5529	5L.46	4S.37
				5-4------	8622	5L.52	4L.30
5-3-------	g	5S.73	3L.01	5-4------	018-4	5L.67	4L.61
5-3-------	4635	5S.48	3S.44	5-4------	b	5L.68	4L.76
5-3-------	6473	5S.26	3S.32	5-4------	027-10	5L.79	4L.61
5-3-------	015-18	5S.21	3L.18	5-4------	8395	5L.82	4L.63
5-3-------	e	5S.16	3S.34				
5-3-------	4873	5S.16	3S.24	5-6------	5622	5S.94	6L.92
5-3-------	4880	5ctr.	3ctr.	5-6------	8818	5S.91	6L.93

TABLE 4.--Continued.

Translocation	Symbol	Breakage points Ratio	Breakage points Ratio	Translocation	Symbol	Breakage points Ratio	Breakage points Ratio
5-6-------	6522	5S.87	6L.70	5-8------	d	5S.55	8L.12
5-6-------	6559	5S.72	6L.09	5-8------	5570	5S.47	8S.35
5-6-------	d	5S.64	6S.89	5-8------	8513	5S.34	8L.24
5-6-------	040-1	5S.48	6S.82	5-8------	c	5S.24	8L.20
5-6-------	f	5S.37	6S.76	5-8------	b	5S.23	8L.23
5-6-------	8590 }	5S.29	6L.25	5-8------	5575	5S.21	8S.22
5-6-------	8593			5-8------	7068	5S.18	8L.18
5-6-------	4933	5S.23	6L.89	5-8------	5777	5S.13	8L.19
5-6-------	6482	5S.22	6S.77	5-8------	017-8	5S.07	8S.25
5-6-------	5765	5S.19	6L.32	5-8------	6402	5S.07	8L.07
5-6-------	5906	5S.15	6L.13	5-8------	A-50	5S.07	8L.11
5-6-------	e	5L.11	6L.60	5-8------	6406	5ctr.	8ctr.
5-6-------	4669	5L.13	6L.40	5-8------	f	5L.02	8S.08
5-6-------	6062	5L.20	6L.78	5-8------	4302-116	5L.06	8S.11
5-6-------	5685	5L.27	6L.20	5-8------	6289	5L.06	8L.54
5-6-------	4934	5L.34	6L.89	5-8------	045-6	5L.08	8L.13
5-6-------	a	5L.35	6L.43	5-8------	002-17	5L.11	8L.28
5-6-------	4666	5L.35	6L.86	5-8------	8997	5L.16	8L.08
5-6-------	8665	5L.58	6L.25·	5-8------	8955	5L.17	8L.83
5-6-------	004-17	5L.60	6L.24	5-8------	014-5	5L.19	8L.18
5-6-------	b	5L.72	6L.21	5-8------	053-4	5L.21	8S.48
5-6-------	8219	5L.76	6S.84	5-8------	4636	5L.23	8L.79
5-6-------	c	5L.81	6L.08	5-8------	g	5L.28	8S.44
5-6-------	8696	5L.89	6S.80	5-8------	007-17	5L.32	8S.47
5-6-------	8379	5L.89	6L.67	5-8------	5866	5L.32	8L.77
				5-8------	7102	5L.48	8S.10
5-7-------	d	5S.63	7S.33	5-8------	030-1	5L.48	8L.78
5-7-------	064-18	5S.61	7S.49	5-8------	a	5L.49	8S.58
5-7-------	061-4	5S.54	7L.30	5-8------	8806	5L.72	8S.59
5-7-------	5143	5S.51	7L.10	5-8------	8796	5L.76	8L.11
5-7-------	e	5S.40	7S.18	5-8------	055-20	5L.81	8L.67
5-7-------	013-3	5S.36	7S.35				
5-7-------	4306-4	5S.32	7L.35	5-9------	068-12	5S.71	9L.86
5-7-------	6372	5S.20	7L.14	5-9------	8854	5S.33	9S.36
5-7-------	8679	5S.09	7S.26	5-9------	022-11	5S.30	9L.27
5-7-------	8491	5ctr.	7ctr.	5-9------	B-91	5S.23	9L.21
5-7-------	062-18	5ctr.	7ctr.	5-9------	B-94	5S.17	9L.27
5-7-------	023-13	5L.12	7L.15	5-9------	6057	5S.15	9S.52
5-7-------	b	5L.18	7S.36	5-9------	8591	5S.09	9L.25
5-7-------	6293	5L.26	7L.63	5-9------	c	5S.07	9L.10
5-7-------	8630	5L.38	7L.24	5-9------	020-7	5ctr.	9ctr.
5-7-------	c	5L.42	7L.72	5-9------	4817	5L.06	9S.07
5-7-------	5179	5L.55	7L.73	5-9------	5614	5L.09	9L.06
5-7-------	a	5L.78	7L.72	5-9------	d	5L.14	9L.10
5-7-------	f	5L.80	7L.85	5-9------	032-8	5L.19	9L.70
5-7-------	8671	5L.96	7L.67	5-9------	7205	5L.21	9L.90
				5-9------	008-18	5L.29	9L.26
5-8-------	8420	5S.90	8L.33	5-9------	044-8	5L.33	9L.55
5-8-------	8746	5S.84	8L.25	5-9------	4790	5L.34	9L.45
5-8-------	8458	5S.68	8L.32	5-9------	8704	5L.35	9L.85
5-8-------	5013	5S.67	8L.59	5-9------	8895	5L.37	9L.11
5-8-------	6612	5S.59	8L.66	5-9------	4305-12	5L.41	9L.64
5-8-------	013-11	5S.59	8S.63	5-9------	4305-22	5L.42	9L.15

TABLE 4.--Continued.

Trans-location	Symbol	Breakage points Ratio	Breakage points Ratio
5-9--------	8936	5L.43	9L.80
5-9--------	e	5L.46	9L.74
5-9--------	4352	5L.48	9L.61
5-9--------	015-10	5L.50	9L.20
5-9--------	013-9	5L.51	9L.82
5-9--------	b	5L.68	9L.44
5-9--------	a	5L.69	9S.17
5-9--------	4871	5L.71	9S.38
5-9--------	8457	5L.78	9S.83
5-9--------	6200	5L.81	9L.71
5-9--------	8386	5L.87	9S.13
5-9--------	5368	5L.87	9L.95
5-10-------	6760	5S.78	10S.40
5-10-------	5355	5S.77	10L.45
5-10-------	5653	5S.76	10L.71
5-10-------	6061	5S.60	10L.57
5-10-------	031-18	5S.58	10S.55
5-10-------	X-57-16	5S.42	10L.42
5-10-------	5679	5S.16	10L.15
5-10-------	6830	5ctr.	10ctr.
5-10-------	b	5L.09	10S.25
5-10-------	5358	5L.10	10L.76
5-10-------	073-6	5L.13	10S.41
5-10-------	4384	5L.13	10L.79
5-10-------	a	5L.14	10S.54
5-10-------	5188	5L.37	10S.65
5-10-------	006-11	5L.49	10L.52
5-10-------	022-20	5L.65	10S.62
5-10-------	7142	5L.73	10L.17
5-10-------	5290	5L.78	10S.49
5-10-------	5688	5L.78	10L.53
5-10-------	8345	5L.87	10S.61
5-10-------	4801	5L.91	10L.23
5-10-------	5557	5L.92	10S.39
6-1--------	8415	6S.82	1L.29
6-1--------	5495	6S.80	1S.25
6-1--------	4986	6S.78	1S.21
6-1--------	d	6S.74	1L.13
6-1--------	6189	6L.17	1S.23
6-1--------	h	6L.17	1L.03
6-1--------	e	6L.21	1S.37
6-1--------	5537	6L.22	1S.31
6-1--------	c	6L.27	1S.25
6-1--------	5013	6L.28	1S.26
6-1--------	4456	6L.30	1L.71
6-1--------	f	6L.42	1L.32
6-1--------	5072	6L.45	1S.17
6-1--------	055-10	6L.48	1S.29
6-1--------	8452	6L.52	1S.80
6-1--------	028-13	6L.54	1S.56

Trans-location	Symbol	Breakage points Ratio	Breakage points Ratio
6-1--------	a	6L.54	1L.20
6-1--------	070-1	6L.58	1L.40
6-1--------	8605 8609 }	6L.59	1S.79
6-1--------	7352	6L.60	1S.40
6-1--------	5077	6L.60	1S.20
6-1--------	7097	6L.62	1S.46
6-1--------	5225	6L.72	1L.61
6-1--------	g	6L.84	1L.16
6-1--------	8650 8658 }	6L.91	1L.79
6-2--------	001-15	6S.87	2S.72
6-2--------	4372	6S.81	2S.37
6-2--------	5419	6S.79	2L.82
6-2--------	8441	6S.79	2L.94
6-2--------	8786	6S.77	2S.90
6-2--------	027-4	6L.11	2S.34
6-2--------	4394	6L.12	2S.91
6-2--------	5472 5473 }	6L.15	2S.25
6-2--------	060-5	6L.18	2S.36
6-2--------	5648	6L.19	2L.25
6-2--------	e	6L.20	2L.18
6-2--------	a	6L.20	2L.28
6-2--------	014-11	6L.20	2L.81
6-2--------	6671	6L.22	2S.22
6-2--------	6931	6L.23	2L.24
6-2--------	c	6L.25	2L.37
6-2--------	008-5	6L.26	2S.40
6-2--------	4717	6L.27	2L.77
6-2--------	"78"	6L.29	2L.24
6-2--------	d	6L.45	2L.41
6-2--------	b	6L.49	2S.69
6-2--------	9002 9009 }	6L.50	2L.57
6-2--------	f	6L.87	2L.79
6-3--------	b	6S.82	3S.73
6-3--------	030-8	6S.81	3S.27
6-3--------	6266	6S.80	3L.85
6-3--------	032-3	6S.78	3S.41
6-3--------	060-4	6L.08	3S.62
6-3--------	8963	6L.14	3S.23
6-3--------	6349	6L.15	3L.10
6-3--------	003-2	6L.15	3L.35
6-3--------	5368	6L.20	3L.22
6-3--------	5201	6L.21	3L.26
6-3--------	8145	6L.26	3L.17
6-3--------	016-17	6L.30	3S.48
6-3--------	a	6L.30	3L.06
6-3--------	055-5	6L.32	3L.16

TABLE 4.--Continued.

Trans-location	Symbol	Breakage points Ratio	Ratio	Trans-location	Symbol	Breakage points Ratio	Ratio
6-3--------	6566	6L.35	3L.41	6-5-------	e	6L.60	5L.11
6-3--------	7162	6L.53	3L.52	6-5-------	8379	6L.67	5L.89
6-3--------	c	6L.54	3S.56	6-5-------	6522	6L.70	5S.87
6-3--------	4349	6L.70	3S.58	6-5-------	6062	6L.78	5L.20
6-3--------	054-12	6L.75	3L.72	6-5-------	4666	6L.86	5L.35
6-3--------	7067	6L.75	3L.07	6-5-------	4933	6L.89	5S.23
6-3--------	d	6L.82	3L.23	6-5-------	4934	6L.89	5L.34
6-3--------	8672	6L.87	3L.47	6-5-------	8818	6L.93	5S.91
6-4--------	003-16	6S.90	4L.50	6-7-------	7036	6S.90	7L.63
6-4--------	7328	6S.89	4S.53	6-7-------	035-3	6S.80	7L.20
6-4--------	5227	6S.84	4S.46	6-7-------	5181	6S.79	7L.86
6-4--------	c	6S.83	4S.33	6-7-------	4964	6S.76	7L.72
6-4--------	4341	6S.81	4S.37	6-7-------	054-6	6L.10	7L.60
6-4--------	7037	6S.77	4L.61	6-7-------	6498	6L.16	7S.48
6-4--------	4447	6L.14	4S.28	6-7-------	4573	6L.22	7L.27
6-4--------	b	6L.16	4S.80	6-7-------	4545	6L.25	7S.73
6-4--------	4461	6L.17	4S.86	6-7-------	7380	6L.29	7L.45
6-4--------	8380	6L.18	4S.47	6-7-------	011-11	6L.29	7L.29
6-4--------	8927	6L.18	4L.70	6-7-------	013-8	6L.31	7L.22
6-4--------	8591 }	6L.24	4L.17	6-7-------	6885	6L.33	7S.58
6-4--------	8593			6-7-------	8143	6L.35	7L.36
6-4--------	055-8	6L.25	4L.29	6-7-------	4337	6L.37	7L.13
6-4--------	8428	6L.28	4L.32	6-7-------	6598	6L.43	7L.61
6-4--------	038.11	6L.29	4L.78	6-7-------	4594	6L.52	7S.67
6-4--------	6623	6L.31	4L.18	6-7-------	027-6	6L.66	7L.97
6-4--------	011-16	6L.33	4S.31	6-7-------	a	6L.73	7L.68
6-4--------	025-12	6L.34	4S.44	6-7-------	7402	6L.97	7L.14
6-4--------	a	6L.43	4L.37				
6-4--------	d	6L.53	4L.49	6-8-------	017-14	6S.92	8L.83
6-4--------	e	6L.56	4S.62	6-8-------	5386	6S.78	8S.83
6-4--------	8339	6L.79	4L.87	6-8-------	058-1	6ctr.	8L.46
6-4--------	8764	6L.90	4L.32	6-8-------	6187	6L.19	8L.51
				6-8-------	6873	6L.21	8L.29
6-5--------	d	6S.89	5S.64	6-8-------	5028	6L.21	8L.31
6-5--------	8219	6S.84	5L.76	6-8-------	c	6L.27	8L.50
6-5--------	040-1	6S.82	5S.48	6-8-------	5605	6L.36	8L.22
6-5--------	8696	6S.80	5L.89	6-8-------	a	6L.41	8L.80
6-5--------	6482	6S.77	5S.22	6-8-------	024-1	6L.42	8L.74
6-5--------	f	6S.76	5S.37	6-8-------	d	6L.51	8L.77
6-5--------	c	6L.08	5L.81	6-8-------	b	6L.79	8S.76
6-5--------	6559	6L.09	5S.72				
6-5--------	5906	6L.13	5S.15	6-9-------	017-14	6S.80	9L.50
6-5--------	5685	6L.20	5L.27	6-9-------	4778	6S.80	9L.30
6-5--------	b	6L.21	5L.72	6-9-------	a	6S.79	9L.40
6-5--------	004-17	6L.24	5L.60	6-9-------	d	6S.73	9L.82
6-5--------	8590 }	6L.25	5S.29	6-9-------	067-6	6S.39	9L.47
6-5--------	8593			6-9-------	5454	6ctr.	9S.75
6-5--------	8665	6L.25	5L.58	6-9-------	84-39	6L.06	9S.73
6-5--------	5765	6L.32	5S.19	6-9-------	b	6L.10	9S.37
6-5--------	4669	6L.40	5L.13	6-9-------	4505	6L.13	9ctr.
6-5--------	5622	6L.42	5S.94	6-9-------	c	6L.15	9L.29
6-5--------	a	6L.43	5L.35	6-9-------	8536	6L.18	9S.81

29

TABLE 4.--Continued.

Trans-location	Symbol	Breakage points Ratio	Breakage points Ratio	Trans-location	Symbol	Breakage points Ratio	Breakage points Ratio
6-9--------	e	6L.18	9L.24	7-1-------	010-12	7L.57	1S.35
6-9--------	8592	6L.19	9S.41	7-1-------	j	7L.61	1L.20
6-9--------	6270	6L.19	9L.28	7-1-------	4891	7L.69	1L.12
6-9--------	018-12	6L.21	9L.25	7-1-------	f	7L.80	1S.72
6-9--------	6019	6L.27	9L.26	7-1-------	4420	7L.90	1L.47
6-9--------	5831	6L.27	9L.30	7-1-------	012-10	7L.90	1L.91
6-9--------	8906	6L.27	9L.59				
6-9--------	043-1	6L.36	9L.36	7-2-------	038-12	7S.68	2L.75
6-9--------	5964	6L.47	9L.83	7-2-------	8465	7S.56	2L.27
6-9--------	8768	6L.89	9S.61	7-2-------	48-74-1	7S.47	2L.22
				7-2-------	c	7S.34	2L.47
6-10-------	f	6S.92	10S.28	7-2-------	8045	7L.06	2S.12
6-10-------	5253	6S.80	10L.41	7-2-------	5144	7L.08	2S.35
6-10-------	5519	6S.75	10L.17	7-2-------	5783	7L.10	2L.66
6-10-------	b	6L.12	10L.29	7-2-------	b	7L.12	2L.37
6-10-------	e	6L.14	10S.43	7-2-------	d	7L.18	2L.16
6-10-------	d	6L.16	10L.29	7-2-------	022-4	7L.24	2S.30
6-10-------	8645	6L.21	10L.28	7-2-------	5279	7L.25	2S.93
6-10-------	8651	6L.27	10L.48	7-2-------	4400	7L.32	2L.24
6-10-------	h	6L.47	10L.87	7-2-------	e	7L.63	2L.82
6-10-------	044-8	6L.48	10L.51	7-2-------	4519	7L.66	2L.65
6-10-------	c	6L.51	10S.36	7-2-------	f	7L.68	2L.76
6-10-------	8904	6L.51	10L.83	7-2-------	8322	7L.74	2L.76
6-10-------	4347	6L.65	10S.81				
6-10-------	4307-12	6L.74	10S.71	7-3-------	6557	7S.59	3L.06
6-10-------	a	6L.75	10L.15	7-3-------	e	7S.56	3L.25
6-10-------	4833	6L.83	10S.78	7-3-------	5724	7ctr.	3ctr.
6-10-------	g	6L.85	10L.20	7-3-------	8541	7ctr.	3ctr.
6-10-------	5780	6L.93	10L.13	7-3-------	b	7L.03	3S.92
				7-3-------	4773	7L.07	3S.11
7-1--------	021-7	7S.56	1S.22	7-3-------	029-3	7L.13	3L.11
7-1--------	4444	7S.50	1S.65	7-3-------	6466	7L.14	3L.36
7-1--------	f	7S.50	1S.10	7-3-------	a	7L.18	3S.25
7-1--------	4405	7S.46	1S.43	7-3-------	004-7	7L.26	3S.38
7-1--------	d	7S.44	1L.81	7-3-------	001-15	7L.30	3S.38
7-1--------	6796	7S.39	1S.40	7-3-------	c	7L.45	3L.46
7-1--------	g	7S.22	1S.79	7-3-------	5955	7L.58	3L.10
7-1--------	b	7S.12	1L.53	7-3-------	5471	7L.58	3L.64
7-1--------	4742	7L.03	1S.95	7-3-------	5378	7L.73	3L.13
7-1--------	e	7L.11	1L.39	7-3-------	4670	7L.76	3S.20
7-1--------	4302	7L.12	1S.15	7-3-------	d	7L.81	3L.64
7-1--------	017-10	7L.12	1L.82	7-3-------	8006	7L.90	3L.88
7-1--------	a	7L.13	1L.28				
7-1--------	5339	7L.14	1L.24	7-4-------	7067	7S.60	4L.17
7-1--------	c	7L.14	1L.39	7-4-------	7108	7S.45	4S.17
7-1--------	5693	7L.18	1L.92	7-4-------	6575	7S.32	4S.38
7-1--------	h	7L.19	1L.46	7-4-------	a	7L.06	4S.32
7-1--------	5871	7L.24	1L.12	7-4-------	008-16	7L.17	4L.86
7-1--------	48-83-15	7L.25	1L.12	7-4-------	052-13	7L.25	4L.27
7-1--------	i	7L.26	1S.31	7-4-------	027-17	7L.31	4L.17
7-1--------	4837	7L.55	1S.73	7-4-------	8374	7L.55	4L.24
7-1--------	A-37	7L.56	1L.10	7-4-------	4483	7L.61	4L.39

TABLE 4.--Continued.

Translocation	Symbol	Breakage points Ratio	Ratio	Translocation	Symbol	Breakage points Ratio	Ratio
7-4	48-40-8	7L.64	4S.32	7-8	016-15	7ctr.	8ctr.
7-4	7347	7L.66	4S.31	7-8	034-17	7L.05	8S.59
7-4	4698	7L.74	4L.08	7-8	5499	7L.05	8L.08
7-4	8103	7L.76	4S.81	7-8	5413	7L.09	8L.77
				7-8	062-16	7L.15	8L.17
7-5	064-18	7S.49	5S.61	7-8	014-17	7L.18	8L.30
7-5	b	7S.36	5L.18	7-8	4536	7L.34	8L.47
7-5	013-3	7S.35	5S.36	7-8	038-8	7L.52	8L.46
7-5	d	7S.33	5S.63	7-8	7149	7L.56	8L.65
7-5	8679	7S.26	5S.09	7-8	5479	7L.70	8S.21
7-5	e	7S.18	5S.40	7-8	021-1	7L.72	8L.49
7-5	8491	7ctr.	5ctr.	7-8	4824	7L.83	8L.25
7-5	062-18	7ctr.	5ctr.	7-8	6427	7L.96	8L.28
7-5	5143	7L.10	5S.51				
7-5	6372	7L.14	5S.20	7-9	b	7S.76	9S.19
7-5	023-13	7L.15	5L.12	7-9	071-1	7S.70	9L.07
7-5	8630	7L.24	5L.38	7-9	8659	7S.55	9S.35
7-5	061-4	7L.30	5S.54	7-9	053-8	7S.51	9L.77
7-5	4306-4	7L.35	5S.32	7-9	5074	7S.48	9L.53
7-5	6293	7L.63	5L.26	7-9	8558	7S.22	9L.16
7-5	8671	7L.67	5L.96	7-9	4363	7ctr.	9ctr.
7-5	c	7L.72	5L.42	7-9	6225	7ctr.	9ctr.
7-5	a	7L.72	5L.78	7-9	8383	7ctr.	9ctr.
7-5	5179	7L.73	5L.55	7-9	6482	7L.01	9S.97
7-5	f	7L.85	5L.80	7-9	7074	7L.03	9S.80
				7-9	c	7L.14	9L.22
7-6	4545	7S.73	6L.25	7-9	8611	7L.23	9L.18
7-6	4594	7S.67	6L.52	7-9	008-16	7L.51	9L.62
7-6	6885	7S.58	6L.33	7-9	4713	7L.60	9ctr.
7-6	6498	7S.48	6L.16	7-9	027-9	7L.61	9S.18
7-6	4337	7L.13	6L.37	7-9	6978	7L.62	9S.83
7-6	7402	7L.14	6L.97	7-9	a	7L.63	9S.07
7-6	035-3	7L.20	6S.80	7-9	008-3	7L.80	9L.85
7-6	013-8	7L.22	6L.31	7-9	032-13	7L.82	9L.88
7-6	4573	7L.27	6L.22	7-9	5381	7L.85	9L.78
7-6	011-11	7L.29	6L.29				
7-6	8143	7L.36	6L.35	7-10	7356	7S.75	10L.88
7-6	7380	7L.45	6L.29	7-10	022-15	7ctr.	10ctr.
7-6	054-6	7L.60	6L.10	7-10	015-12	7ctr.	10ctr.
7-6	6598	7L.61	6L.43	7-10	019-3	7L.17	10L.47
7-6	7036	7L.63	6S.90	7-10	a	7L.23	10L.06
7-6	a	7L.68	6L.73	7-10	4356	7L.70	10S.37
7-6	4964	7L.72	6S.76	7-10	4422	7L.79	10ctr.
7-6	5181	7L.86	6S.79	8-1	5704	8S.67	1L.96
7-6	027-6	7L.97	6L.66	8-1	a	8S.52	1L.41
7-8	8346	7S.49	8S.30	8-1	6591	8S.43	1S.18
7-8	5828	7S.31	8L.10	8-1	5588	8S.32	1S.10
7-8	6531	}7ctr.	8ctr.	8-1	5634	8S.28	1L.08
7-8	6981			8-1	4676	8S.06	1L.04
7-8	8580	}7ctr.	8ctr.	8-1	055-23	8ctr.	1ctr.
7-8	8583			8-1	064-13	8ctr.	1ctr.
7-8	004-3	7ctr.	8ctr.	8-1	8683	8ctr.	1L.11

TABLE 4.--Continued.

Trans-location	Symbol	Breakage points Ratio	Ratio	Trans-location	Symbol	Breakage points Ratio	Ratio
8-1-------	4748	8L.15	1L.12	8-3------	043-14	8S.40	3L.02
8-1-------	5619	8L.16	1L.07	8-3------	015-17	8S.13	3S.12
8-1-------	8640	8L.16	1L.11	8-3------	B-31	8S.06	3L.63
8-1-------	008-17	8L.20	1S.16	8-3------	4303-12	8ctr.	3ctr.
8-1-------	4685	8L.21	1S.20	8-3------	4872	8ctr.	3ctr.
8-1-------	7509	8L.21	1L.28	8-3------	5583	8L.05	3L.06
8-1-------	5752	8L.23	1L.36	8-3------	f	8L.10	3L.08
8-1-------	5821	8L.31	1L.65	8-3------	007-8	8L.11	3L.14
8-1-------	020-19	8L.38	1L.11	8-3------	5830	8L.13	3S.06
8-1-------	8919	8L.44	1S.53	8-3------	8666		
8-1-------	6697	8L.52	1L.89	8-3------	8667	} 8L.14	3S.30
8-1-------	5384	8L.59	1L.10	8-3------	8670		
8-1-------	036-4	8L.59	1L.18	8-3------	6439	8L.15	3S.30
8-1-------	4307	8L.61	1S.42	8-3------	8023	8L.16	3L.18
8-1-------	001-13	8L.67	1S.39	8-3------	g	8L.19	3L.12
8-1-------	5910	8L.67	1L.93	8-3------	e	8L.21	3S.36
8-1-------	6766	8L.77	1L.54	8-3------	b	8L.23	3L.16
8-1-------	005-7	8L.78	1L.22	8-3------	012-10	8L.23	3L.39
8-1-------	026-2	8L.80	1L.49	8-3------	4626	8L.31	3S.30
8-1-------	b	8L.82	1L.59	8-3------	4874	8L.32	3L.28
8-1-------	c	8L.89	1L.94	8-3------	6261	8L.40	3L.49
				8-3------	024-11	8L.49	3S.65
8-2-------	003-5	8S.72	2L.19	8-3------	a	8L.61	3L.41
8-2-------	g	8S.71	2L.71	8-3------	6373	8L.68	3S.53
8-2-------	5484	8S.58	2L.24	8-3------	7362	8L.69	3L.07
8-2-------	031-7	8S.44	2L.30	8-3------	4340	8L.72	3L.88
8-2-------	5454	8S.39	2L.21	8-3------	4301-39	8L.82	3L.92
8-2-------	c	8S.11	2S.15	8-3------	c	8L.85	3S.23
8-2-------	f	8ctr.	2ctr.				
8-2-------	006-10	8ctr.	2ctr.	8-4------	057-21	8S.49	4L.83
8-2-------	8376	8L.03	2L.95	8-4------	016-6	8S.45	4L.42
8-2-------	d	8L.10	2L.05	8-4------	6063	8L.05	4S.02
8-2-------	e	8L.10	2L.07	8-4------	004-12	8L.10	4L.10
8-2-------	8428	8L.10	2L.16	8-4------	4356	8L.14	4S.40
8-2-------	4414	8L.14	2L.12	8-4------	b	8L.16	4S.18
8-2-------	7069	8L.14	2L.13	8-4------	a	8L.19	4S.59
8-2-------	b	8L.18	2L.20	8-4------	6363	8L.30	4L.76
8-2-------	h	8L.22	2L.23	8-4------	8603		
8-2-------	062-15	8L.26	2L.70	8-4------	8606		
8-2-------	011-20	8L.28	2S.58	8-4------	8607	} 8L.35	4S.42
8-2-------	"84"	8L.30	2L.32	8-4------	8608		
8-2-------	8458	8L.32	2S.22	8-4------	036-16	8L.69	4S.66
8-2-------	051-15	8L.48	2L.62	8-4------	5339	8L.71	4S.22
8-2-------	037-5	8L.54	2L.95	8-4------	6926	8L.71	4L.60
8-2-------	013-17	8L.61	2S.89	8-4------	8456	8L.75	4S.22
8-2-------	4711	8L.67	2S.86	8-4------	8987	8L.76	4S.58
8-2-------	48-45-6	8L.68	2L.84	8-4------	8677	8L.79	4S.24
8-2-------	051-7	8L.74	2L.83	8-4------	8004	8L.84	4S.27
				8-4------	8163	8L.86	4S.17
8-3-------	5558	8S.74	3S.26	8-4------	5412	8L.95	4L.59
8-3-------	8350	8S.60	3L.75				
8-3-------	8367	8S.52	3S.28	8-5------	013-11	8S.63	5S.59
8-3-------	h	8S.46	3L.53	8-5------	8806	8S.59	5L.72

TABLE 4.--Continued.

Trans-location	Symbol	Breakage points Ratio	Ratio	Trans-location	Symbol	Breakage points Ratio	Ratio
8-5-------	a	8S.58	5L.49	8-7------	8580	} 8ctr.	7ctr.
8-5-------	053-4	8S.48	5L.21	8-7------	8583		
8-5-------	007-17	8S.47	5L.32	8-7------	004-3	} 8ctr.	7ctr.
8-5-------	g	8S.44	5L.28	8-7------	016-15		
8-5-------	5570	8S.35	5S.47	8-7------	5499	8L.08	7L.05
8-5-------	017-8	8S.25	5S.09	8-7------	5828	8L.10	7S.31
8-5-------	5575	8S.22	5S.21	8-7------	062-16	8L.17	7L.15
8-5-------	4302-116	8S.11	5L.06	8-7------	4824	8L.25	7L.83
8-5-------	7102	8S.10	5L.48	8-7------	6427	8L.28	7L.96
8-5-------	f	8S.08	5L.02	8-7------	014-17	8L.30	7L.18
8-5-------	6406	8ctr.	5ctr.	8-7------	038-8	8L.46	7L.52
8-5-------	6402	8L.07	5S.07	8-7------	4536	8L.47	7L.34
8-5-------	8997	8L.08	5L.16	8-7------	021-1	8L.49	7L.72
8-5-------	A-50	8L.11	5S.07	8-7------	7149	8L.65	7L.56
8-5-------	8796	8L.11	5L.76	8-7------	5413	8L.77	7L.09
8-5-------	d	8L.12	5S.55				
8-5-------	045-6	8L.13	5L.08	8-9------	b	8S.67	9L.75
8-5-------	7067	8L.18	5S.18	8-9------	8661	8S.66	9L.92
8-5-------	014-5	8L.18	5L.19	8-9------	4643	8S.37	9L.11
8-5-------	5777	8L.19	5S.13	8-9------	c	8ctr.	9ctr.
8-5-------	c	8L.20	5S.24	8-9------	8525	8L.06	9S.63
8-5-------	b	8L.23	5S.23	8-9------	5391	8L.07	9S.33
8-5-------	8513	8L.24	5S.34	8-9------	d	8L.09	9S.16
8-5-------	8746	8L.25	5S.84	8-9------	a	8L.13	9L.38
8-5-------	002-17	8L.28	5L.11	8-9------	8951	8L.13	9L.77
8-5-------	8458	8L.32	5S.68	8-9------	034-11	8L.14	9L.18
8-5-------	8420	8L.33	5S.90	8-9------	043-6	8L.17	9S.34
8-5-------	6289	8L.54	5L.06	8-9------	e	8L.32	9L.25
8-5-------	5013	8L.59	5S.67	8-9------	6673	8L.35	9S.31
8-5-------	6612	8L.66	5S.59	8-9------	4775	8L.42	9L.68
8-5-------	055-20	8L.67	5L.81	8-9------	018-12	8L.52	9S.49
8-5-------	5866	8L.77	5L.32	8-9------	4593	8L.69	9L.65
8-5-------	030-1	8L.78	5L.48	8-9------	6921	8L.85	9L.15
8-5-------	4636	8L.79	5L.23	8-9------	5300	8L.85	9S.43
8-5-------	8955	8L.83	5L.17	8-9------	4453	8L.86	9S.68
8-6-------	5386	8S.83	6S.78	8-10-----	023-15	8S.42	10S.31
8-6-------	b	8S.76	6L.79	8-10-----	b	8ctr.	10ctr.
8-6-------	5605	8L.22	6L.36	8-10-----	5585	8ctr.	10ctr.
8-6-------	6873	8L.29	6L.21	8-10-----	6653	8L.04	10L.06
8-6-------	5028	8L.31	6L.21	8-10-----	K-7	8L.07	10S.22
8-6-------	058-1	8L.46	6ctr.	8-10-----	9020	8L.13	10S.50
8-6-------	c	8L.50	6L.27	8-10-----	6488	8L.14	10S.34
8-6-------	6187	8L.51	6L.19	8-10-----	5287	8L.17	10S.33
8-6-------	024-1	8L.74	6L.42	8-10-----	034-19	8L.24	10L.28
8-6-------	d	8L.77	6L.51	8-10-----	001-5	8L.30	10S.57
8-6-------	a	8L.80	6L.41	8-10-----	d	8L.39	10L.16
8-6-------	017-14	8L.83	6S.92	8-10-----	c	8L.41	10S.56
				8-10-----	6128	8L.43	10S.49
8-7-------	034-17	8S.59	7L.05	8-10-----	a	8L.48	10S.48
8-7-------	8349	8S.30	7S.49	8-10-----	5944	8L.75	10L.40
8-7-------	5479	8S.21	7L.70	8-10-----	e	8L.84	10S.37
8-7-------	6531	} 8ctr.	7ctr.	9-1------	035-10	9S.67	1L.89
8-7-------	6981						

33

TABLE 4.--Continued.

Translocation	Symbol	Breakage points Ratio	Ratio	Translocation	Symbol	Breakage points Ratio	Ratio
9-1-------	4997	9S.28	1L.37	9-3------	030-2	9L.30	3S.39
9-1-------	7535	9S.27	1S.33	9-3------	h	9L.33	3L.09
9-1-------	4995	9S.20	1L.19	9-3------	8465	9L.41	3S.27
9-1-------	4398	9S.19	1L.51	9-3------	4727	9L.42	3L.54
9-1-------	5622	9L.12	1L.10	9-3------	5285	9L.49	3L.51
9-1-------	024-7	9L.13	1S.71	9-3------	b	9L.53	3L.48
9-1-------	8389	9L.13	1L.74	9-3------	4963	9L.57	3L.76
9-1-------	a	9L.15	1S.13	9-3------	5643	9L.64	3S.55
9-1-------	8918	9L.20	1S.21	9-3------	7041	9L.70	3S.59
9-1-------	c	9L.22	1S.48	9-3------	064-11	9L.81	3L.17
9-1-------	8886 }	9L.23	1L.33	9-3------	054-18	9L.82	3S.88
9-1-------	8889 }			9-3------	8032	9L.96	3S.26
9-1-------	48-34-2	9L.24	1S.51				
9-1-------	8001	9L.24	1S.51	9-4------	6504	9S.83	4L.09
9-1-------	8460	9L.24	1S.13	9-4------	6222	9S.68	4L.03
9-1-------	d	9L.25	1L.42	9-4------	5657	9S.25	4L.33
9-1-------	8265	9L.27	1S.58	9-4------	8636 }	9S.09	4L.94
9-1-------	8129	9L.27	1S.53	9-4------	8649 }		
9-1-------	8302	9L.29	1S.55	9-4------	d	9L.17	4L.12
9-1-------	6328	9L.40	1L.79	9-4------	5918	9L.18	4S.24
9-1-------	6762	9L.53	1S.16	9-4------	f	9L.18	4L.55
9-1-------	b	9L.60	1L.50	9-4------	e	9L.26	4S.53
9-1-------	8824	9L.83	1S.06	9-4------	004-7	9L.26	4L.28
9-1-------	9021	9L.83	1L.26	9-4------	g	9L.27	4S.27
				9-4------	5828	9L.27	4L.61
9-2-------	5831	9S.72	2L.63	9-4------	c	9L.29	4L.82
9-2-------	062-11	9S.53	2L.21	9-4------	b	9L.29	4L.90
9-2-------	c	9S.33	2S.49	9-4------	4304-82	9L.37	4S.22
9-2-------	6656	9S.31	2L.32	9-4------	4373	9L.39	4L.29
9-2-------	5257	9L.20	2L.28	9-4------	5884	9L.49	4L.40
9-2-------	b	9L.22	2S.18	9-4------	4307-12	9L.55	4S.48
9-2-------	5711	9L.23	2S.24	9-4------	a	9L.58	4L.16
9-2-------	055-14	9L.27	2S.28	9-4------	5788	9L.82	4L.72
9-2-------	d	9L.27	2L.83	9-4------	5574	9L.87	4L.80
9-2-------	a	9L.58	2S.36	9-4------	4333	9L.93	4L.41
9-2-------	7096	9L.66	2S.57				
9-2-------	5208	9L.68	2L.76	9-5------	8457	9S.83	5L.78
9-2-------	(2-8a)	9L.72	2S.27	9-5------	6057	9S.52	5S.15
				9-5------	4871	9S.38	5L.71
9-3-------	f	9S.69	3L.63	9-5------	8854	9S.36	5S.33
9-3-------	6722	9S.66	3S.66	9-5------	a	9S.17	5L.69
9-3-------	034-11	9S.36	3L.46	9-5------	8386	9S.13	5L.87
9-3-------	5775	9S.24	3L.09	9-5------	4817	9S.07	5L.06
9-3-------	020-5	9ctr.	3ctr.	9-5------	020-7	9ctr.	5ctr.
9-3-------	c	9L.12	3L.09	9-5------	5614	9L.06	5L.09
9-3-------	8447	9L.14	3S.44	9-5------	c	9L.10	5S.07
9-3-------	g	9L.14	3L.40	9-5------	d	9L.10	5L.14
9-3-------	a	9L.16	3L.11	9-5------	8895	9L.11	5L.37
9-3-------	8562 }			9-5------	4305-22	9L.15	5L.42
9-3-------	8568 }	9L.22	3L.65	9-5------	015-10	9L.20	5L.50
9-3-------	8572 }			9-5------	B-91	9L.21	5S.23
9-3-------	8574 }			9-5------	8591	9L.25	5S.09
9-3-------	d	9L.26	3L.13	9-5------	008-18	9L.26	5L.29

022-11	9L.27	5S.30	9-7------	c	9L.22	7L.I4
B-94	9L.27	5S.17	9-7------	5074	9L.53	7S.48
b	9L.44	5L.68	9-7------	008-16	9L.62	7L.51
4790	9L.45	5L.34	9-7------	053-8	9L.77	7S.51
044-8	9L.55	5L.33	9-7------	5381	9L.78	7L.85
4352	9L.61	5L.48	9-7------	008-3	9L.85	7L.80
4305-12	9L.64	5L.41	9-7------	032-13	9L.88	7L.82
032-8	9L.70	5L.19				
6200	9L.71	5L.81	9-8------	4453	9S.68	8L.86
e	9L.74	5L.46	9-8------	8525	9S.63	8L.06
8936	9L.80	5L.43	9-8------	018-12	9S.49	8L.52
013-9	9L.82	5L.51	9-8------	5300	9S.43	8L.85
8704	9L.85	5L.35	9-8------	043-6	9S.34	8L.17
068-12	9L.86	5S.71	9-8------	5391	9S.33	8L.07
7205	9L.90	5L.21	9-8------	6673	9S.31	8L.35
5368	9L.95	5L.87	9-8------	d	9S.16	8L.09
			9-8------	c	9ctr.	8ctr.
8536	9S.81	6L.18	9-8------	4346	9L.11	8S.37
5454	9S.75	6ctr.	9-8------	6921	9L.15	8L.85
8439	9S.73	6L.06	9-8------	034-11	9L.18	8L.14
8768	9S.61	6L.89	9-8------	e	9L.25	8L.32
8592	9S.41	6L.19	9-8------	a	9L.38	8L.13
b	9S.37	6L.10	9-8------	4593	9L.65	8L.69
4505	9ctr.	6L.13	9-8------	4775	9L.68	8L.42
e	9L.24	6L.18	9-8------	b	9L.75	8S.67
018-12	9L.25	6L.21	9-8------	8951	9L.77	8L.13
6019	9L.26	6L.27	9-8------	8661	9L.92	8S.66
6270	9L.28	6L.19				
c	9L.29	6L.15	9-10-----	059-10	9S.31	10L.53
4778	9L.30	6S.80	9-10-----	8630	9S.28	10L.37
5831	9L.30	6L.27	9-10-----	B-49	9S.21	10L.60
043-1	9L.36	6L.36	9-10-----	b	9S.13	10S.40
a	9L.40	6S.79	9-10-----	a	9L.14	10L.92
067-6	9L.47	6S.39	9-10-----	A-75	9L.24	10L.30
017-14	9L.50	6S.80	9-10-----	4303-9	9L.26	10S.44
8906	9L.59	6L.27	9-10-----	5488	9L.57	10L.89
d	9L.82	6S.73	9-10-----	041-4	9L.67	10L.92
5964	9L.83	6L.47	9-10-----	041-6	9L.70	10L.90
			9-10-----	7103	9L.73	10L.88
6482	9S.97	7L.01				
6978	9S.83	7L.62	10-1-----	015-9	10S.46	1L.67
7074	9S.80	7L.03	10-1-----	b	10S.39	1L.19
8659	9S.35	7S.55	10-1-----	4885	10ctr.	1ctr.
b	9S.19	7S.76	10-1-----	007-19	10L.08	1S.07
027-9	9S.18	7L.61	10-1-----	g	10L.21	1L.80
a	9S.07	7L.63	10-1-----	B-29	10L.26	1L.93
4363	9ctr.	7ctr.	10-1-----	f	10L.30	1S.04
6225	9ctr.	7ctr.	10-1-----	e	10L.31	1L.16
8383	9ctr.	7ctr.	10-1-----	a	10L.33	1L.29
4713	9ctr.	7L.60	10-1-----	8770	10L.38	1L.09
071-1	9L.07	7S.70	10-1-----	001-3	10L.48	1L.86
8558	9L.16	7S.22	10-1-----	8375	10L.64	1L.69
8611	9L.18	7L.23	10-1-----	d	10L.68	1L.50

TABLE 4.--Continued.

Trans-location	Symbol	Breakage points Ratio	Ratio	Trans-location	Symbol	Breakage points Ratio	Ratio
10-1------	5273	10L.69	1L.17	10-5-----	031-18	10S.55	5S.58
10-1------	c	10L.74	1L.43	10-5-----	a	10S.54	5L.14
10-1------	8465	}10L.76	1L.45	10-5-----	5290	10S.49	5L.78
10-1------	8491			10-5-----	073-6	10S.41	5L.13
10-1------	068-14	10L.79	1L.16	10-5-----	6760	10S.40	5S.78
				10-5-----	5557	10S.39	5L.92
10-2------	I-3	10S.40	2L.30	10-5-----	b	10S.25	5L.09
10-2------	5561	10S.16	2L.35	10-5-----	6830	10ctr.	5ctr.
10-2------	5830	10L.12	2L.12	10-5-----	5679	10L.15	5S.16
10-2------	4484	10L.14	2S.09	10-5-----	7142	10L.17	5L.73
10-2------	035-2	10L.19	2L.85	10-5-----	4801	10L.23	5L.91
10-2------	8219	10L.35	2L.50	10-5-----	X-57-16	10L.42	5S.42
10-2------	043-10	10L.40	2S.89	10-5-----	5355	10L.45	5S.77
10-2------	a	10L.55	2L.16	10-5-----	006-11	10L.52	5L.49
10-2------	5651	10L.62	2S.71	10-5-----	5688	10L.53	5L.78
10-2------	063-1	10L.71	2L.19	10-5-----	6061	10L.67	5S.60
10-2------	b	10L.75	2S.50	10-5-----	5653	10L.71	5S.76
10-2------	8864	10L.76	2S.10	10-5-----	5358	10L.76	5L.10
10-2------	011-9	10L.76	2L.39	10-5-----	4384	10L.79	5L.13
10-2------	6853	10L.86	2L.79				
				10-6-----	4347	10S.81	6L.65
10-3------	8412	10S.36	3S.39	10-6-----	4833	10S.78	6L.83
10-3------	8349	10ctr.	3S.38	10-6-----	4307-12	10S.71	6L.74
10-3------	a	10L.22	3L.16	10-6-----	e	10S.43	6L.14
10-3------	5892	10L.25	3S.17	10-6-----	c	10S.36	6L.51
10-3------	b	10L.27	3L.19	10-6-----	f	10S.28	6S.92
10-3------	4382	10L.29	3S.38	10-6-----	5780	10L.13	6L.93
10-3------	4383	10L.30	3S.23	10-6-----	a	10L.15	6L.75
10-3------	c	10L.30	3L.22	10-6-----	5519	10L.17	6S.75
10-3------	7464	10L.60	3S.47	10-6-----	g	10L.20	6L.85
10-3------	036-15	10L.64	3L.48	10-6-----	8645	10L.28	6L.21
10-3------	063-1	10L.71	3S.25	10-6-----	8651	10L.28	6L.27
10-3------	044-10	10L.72	3L.77	10-6-----	b	10L.29	6L.12
10-3------	6691	10L.87	3L.30	10-6-----	d	10L.29	6L.16
				10-6-----	5253	10L.41	6S.80
10-4------	073-8	10S.74	4L.41	10-6-----	044-8	10L.51	6L.48
10-4------	057-14	10S.48	4L.56	10-6-----	8904	10L.83	6L.51
10-4------	8541	10ctr.	4S.45	10-6-----	h	10L.87	6L.47
10-4------	6662	10L.03	4L.04				
10-4------	e	10L.14	4L.14	10-7-----	4356	10S.37	7L.70
10-4------	f	10L.14	4L.94	10-7-----	022-15	10ctr.	7ctr.
10-4------	c	10L.18	4S.64	10-7-----	015-12	10ctr.	7ctr.
10-4------	024-16	10L.18	4L.75	10-7-----	4422	10ctr.	7L.79
10-4------	021-5	10L.33	4L.34	10-7-----	a	10L.06	7L.23
10-4------	d	10L.36	4S.36	10-7-----	019-3	10L.47	7L.17
10-4------	6587	10L.51	4L.55	10-7-----	7356	10L.88	7S.75
10-4------	b	10L.60	4L.15				
10-4------	9028	}10L.89	4S.57	10-8-----	001-5	10S.57	8L.30
10-4------	9029			10-8-----	c	10S.56	8L.41
				10-8-----	9020	10S.50	8L.13
10-5------	5188	10S.65	5L.37	10-8-----	a	10S.48	8L.48
10-5------	022-20	10S.62	5L.65	10-8-----	6128	10S.47	8L.43
10-5------	8345	10S.61	5L.87	10-8-----	e	10S.37	8L.84

TABLE 4.--Continued.

Trans-location	Symbol	Breakage points		Trans-location	Symbol	Breakage points	
		Ratio	Ratio			Ratio	Ratio
10-8------	6488	10S.34	8L.14	10-9-----	b	10S.40	9S.13
10-8------	5287	10S.33	8L.17	10-9-----	A-75	10L.30	9L.24
10-8------	023-15	10S.31	8S.42	10-9-----	8630	10L.37	9S.28
10-8------	K-7	10S.22	8L.07	10-9-----	059-10	10L.53	9S.31
10-8------	b	10ctr.	8ctr.	10-9-----	B-49	10L.60	9S.21
10-8------	5585	10ctr.	8ctr.	10-9-----	7103	10L.88	9L.73
10-8------	6653	10L.06	8L.04	10-9-----	5488	10L.89	9L.57
10-8------	d	10L.16	8L.39	10-9-----	041-6	10L.90	9L.70
10-8------	034-19	10L.28	8L.24	10-9-----	a	10L.92	9L.14
10-8------	5944	10L.40	8L.75	10-9-----	041-4	10L.92	9L.67
10-9------	4303-9	10S.44	9L.26				

TABLE 5.--Breakage point positions and densities for all chromosomes, and for chromosome arms and sections of these arms, in the total population of translocations.

Chromosome threads involved	Mean break position	Total length	Density of breaks, ob./ex., in--				
			Section 1μ-10μ	Section 10μ-20μ	Section 20μ-30μ	Section 30μ-40μ	Section 40μ-50μ
1-10--------------	0.410	1.000	1.432	0.841	0.752	0.422	0.403
1S-10S------------	.418	.966	1.331	.744	.535	.388	--
1L-10L------------	.406	1.020	1.562	.932	.835	.450	.405
1S----------------	.389	.851	1.630	.98	.709	.462	--
1L----------------	.400	.831	1.787	1.074	.812	.585	.598
2S----------------	.447	.771	1.182	1.035	.776	--	--
2L----------------	.427	1.002	1.405	1.080	.897	.464	--
3S----------------	.381	1.054	1.408	.647	--	--	--
3L----------------	.347	.830	2.032	.84	.775	.485	--
4S----------------	.362	.930	1.695	.556	--	--	--
4L----------------	.429	.850	1.370	.954	.954	.541	--
5S----------------	.405	1.197	1.487	.888	.489	--	--
5L----------------	.413	1.243	1.438	1.046	.806	.267	--
6S----------------	.802	.940	.978	1.800	--	--	--
6L----------------	.393	1.110	1.645	1.153	.54	.504	--
7S----------------	.394	.992	1.274	--	--	--	--
7L----------------	.413	.930	1.600	.524	1.163	.334	--
8S----------------	.360	1.310	1.130	--	--	--	--
8L----------------	.387	1.180	1.615	.867	.889	.413	--
9S----------------	.391	1.010	1.232	.677	--	--	--
9L----------------	.426	1.280	1.463	.785	.687	--	--
10S---------------	.415	1.110	.981	--	--	--	--
10L---------------	.440	.920	1.212	.772	.736	--	--

TABLE 6.--Breakage point positions and densities for 3 subpopulations of translocations produced by ionizing radiation treatments.

Chromosome threads involved	Breakage positions and densities of chromosomes of radiation-treated--					
	Seed--15,000 r. X-ray		Seed--Bikini Lot 3		Pollen--1,000 r. X-ray	
	Mean break position	Density of breaks, ob./ex.	Mean break position	Density of breaks, ob./ex.	Mean break position	Density of breaks, ob./ex.
1-10------------	0.408	1.000	0.418	1.000	0.397	1.000
1S-10S----------	.389	1.031	.416	1.050	.401	.990
1L-10L----------	.422	.982	.421	.972	.392	1.007
1S--------------	.360	.624	.440	.758	.359	.767
1L--------------	.373	.591	.376	1.193	.324	.736
2S--------------	.434	.591	.483	.920	.395	1.310
2L--------------	.408	.766	.526	.937	.379	1.190
3S--------------	.272	1.576	.407	1.607	.364	.986
3L--------------	.364	.808	.403	.786	.333	.679
4S--------------	.337	.776	.320	1.037	.397	.724
4L--------------	.417	.710	.428	.841	.436	1.046
5S--------------	.285	1.364	.459	1.418	.487	1.191
5L--------------	.389	1.031	.412	.882	.382	1.285
6S--------------	.784	1.461	.800	.813	.789	.944
6L--------------	.350	1.190	.398	1.250	.317	.864
7S--------------	.515	.793	.297	1.567	.411	.806
7L--------------	.452	1.352	.478	.778	.367	1.068
8S--------------	.330	.882	.297	1.030	.390	1.555
8L--------------	.438	1.140	.451	1.420	.401	1.246
9S--------------	.409	1.470	.370	1.020	.360	.660
9L--------------	.580	1.195	.354	1.164	.483	1.138
10S-------------	.290	1.020	.472	.990	.460	1.147
10L-------------	.353	.950	.350	.637	.506	.903

TABLE 7.--Breakage points for inversions isolated from various ionizing radiation treatments.

Symbol	Chromosome	Breakage points		Symbol	Chromosome	Breakage points	
		Ratio	Ratio			Ratio	Ratio
4345-------	I1	S.85	L.56	5876------	I4	L.24	L.66
4932-------	I1	S.41	L.35	5887------	I4	L.16	L.56
5357-------	I1	S.19	L.44	5965------	I4	L.19	L.66
5517-------	I1	S.37	L.05	6748------	I4	L.16	L.61
5828-------	I1	S.82	L.46				
6760-------	I1	S.88	L.67	5068------	I5	S.21	L.75
4305-25----	I1	L.71	L.95	5431------	I5	S.80	L.91
5083-------	I1	L.70	L.87	5583------	I5	S.09	L.62
5131-------	I1	L.46	L.82	5787------	I5	S.78	L.63
				6329------	I5	S.41	L.84
4333-------	I2	S.88	L.50	7088------	I5	S.42	L.63
5334-------	I2	S.64	L.96	8611------			
5342-------	I2	S.76	L.86	8623------ }	I5	S.67	L.69
5407-------	I2	S.91	L.50	8624------			
7006-------	I2	S.81	L.85	002-16----	I5	S.51	L.67
8541-------	I2	S.36	L.82	4305-22---	I5	L.26	L.81
8865-------	I2	S.06	L.05	8463------	I5	L.65	L.87
48-51-4----	I2	S.73	L.70				
003-20-----	I2	S.38	L.44	6669------	I6	S.38	L.92
005-19-----	I2	S.69	L.58	8452------	I6	S.77	L.33
4629-------	I2	L.26	L.67	8604------	I6	S.85	L.32
5392-------	I2	L.13	L.51	014-11----	I6	S.80	L.20
6027-------	I2	L.24	L.76				
6597-------	I2	L.08	L.22	5250------	I7	S.51	L.72
				020-15----	I7	S.89	L.93
6131-------	I3	S.26	L.09	5262------	I7	L.11	L.79
6839-------	I3	S.72	L.42	5803------	I7	L.17	L.61
8583-------	I3	S.55	L.82	6482------	I7	L.31	L.66
8664-------	I3	S.39	L.54	8540------			
025-6------	I3	S.46	L.80	8544------ }	I7	L.12	L.92
5272-------	I3	L.19	L.72	8545------			
6784-------	I3	L.21	L.70				
				002-12----	I8	S.51	L.77
4791-------	I4	S.89	L.72	5949------	I8	L.10	L.42
6323-------	I4	S.24	L.39				
4301-------	I4	L.03	L.38	5570------	I10	S.57	L.86

Lightning Source UK Ltd.
Milton Keynes UK
UKHW021146061218
333419UK00013B/2115/P